Springer Complexity

Springer Complexity is an interdisciplinary program publishing the best research and academic-level teaching on both fundamental and applied aspects of complex systems—cutting across all traditional disciplines of the natural and life sciences, engineering, economics, medicine, neuroscience, social and computer science.

Complex Systems are systems that comprise many interacting parts with the ability to generate a new quality of macroscopic collective behavior the manifestations of which are the spontaneous formation of distinctive temporal, spatial or functional structures. Models of such systems can be successfully mapped onto quite diverse "real-life" situations like the climate, the coherent emission of light from lasers, chemical reaction-diffusion systems, biological cellular networks, the dynamics of stock markets and of the internet, earthquake statistics and prediction, freeway traffic, the human brain, or the formation of opinions in social systems, to name just some of the popular applications.

Although their scope and methodologies overlap somewhat, one can distinguish the following main concepts and tools: self-organization, nonlinear dynamics, synergetics, turbulence, dynamical systems, catastrophes, instabilities, stochastic processes, chaos, graphs and networks, cellular automata, adaptive systems, genetic algorithms and computational intelligence.

The three major book publication platforms of the Springer Complexity program are the monograph series "Understanding Complex Systems" focusing on the various applications of complexity, the "Springer Series in Synergetics", which is devoted to the quantitative theoretical and methodological foundations, and the "Springer Briefs in Complexity" which are concise and topical working reports, case studies, surveys, essays and lecture notes of relevance to the field. In addition to the books in these two core series, the program also incorporates individual titles ranging from textbooks to major reference works.

Indexed by SCOPUS, INSPEC, zbMATH, SCImago.

Series Editors

Understanding Complex Systems

Founding Editor: Scott Kelso

More information about this series at http://www.springer.com/series/5394

Oscar M. Granados · José R. Nicolás-Carlock
Editors

Corruption Networks

Concepts and Applications

 Springer

Editors
Oscar M. Granados
Department of Economics
and International Trade
Universidad Jorge Tadeo Lozano
Bogotá, Colombia

José R. Nicolás-Carlock
Institute of Legal Research
National Autonomous University of Mexico
Mexico City, Mexico

ISSN 1860-0832   ISSN 1860-0840   (electronic)
Understanding Complex Systems
ISBN 978-3-030-81486-1   ISBN 978-3-030-81484-7   (eBook)
https://doi.org/10.1007/978-3-030-81484-7

This Springer imprint is published by the registered company Springer Nature Switzerland AG
The registered company address is: Gewerbestrasse 11, 6330 Cham, Switzerland

For next generations free of corruption

Foreword

This book has brought together a diverse set of innovative scholars from political science, sociology, economics, physics, and mathematics using different methods ranging from qualitative to complexity science. A common thread among the contributions in this book the reader holds in her/his hands is that the network representations of social relations underlying corrupt transactions are crucial for understanding and countering the phenomenon. Corruption is rarely an isolated phenomenon representing an outlier against the background of generalized integrity, and, typically, various forms and manifestations of corruption are enabled by trust, mutual monitoring, and sanctioning, facilitated by a network of relationships. Network perspectives can be traced back to early efforts to define corruption as a transactional feature inherent in public service provision, government contracting, tax collection, or natural resource extraction. Additionally, organized crime research has long recognized the networked nature of corruption and how personal connections underpin the corrupt governance of organizations and markets. Thus, there is a growing recognition in different disciplines that corruption typically presents collective action dilemmas, and the sustainability of corrupt societal equilibria is maintained by norms and self-reinforcing expectations. Network analysis of corrupt phenomena holds the key to better understanding their drivers and designing effective anticorruption policies. Furthermore, these perspectives are indispensable as they allow for linking individual corrupt relationships and transactions to meso and macro behavior. Even though research and public policy advice have been interested in macro-level analysis and predominantly focused on fighting corruption on the country level, they often lacked the micro-foundations leading to incomplete conclusions and ill-informed policy recipes. In the last decade, this situation has changed with growing attention paid to local and sectoral corrupt phenomena and the tailored tools effectively fighting corruption in context. However, the disconnect between the micro and macro has neglected, calling for novel methods and approaches capable of filling this void. Network analysis of

corruption dares both to address the micro-phenomena in its context but also aims
to draw macro-relevant conclusions, hence it could be the method, at least partially,
which fills this void.

Vienna, Austria   Mihály Fazekas
May 2021

Preface

According to the United Nations, corruption is a transnational phenomenon that affects societies in several ways on their political, economic, ecological, and social fronts, causing unstable democracies, unfair market competition, harming natural environments, and fostering human rights violations. As such, corruption is a complex, ubiquitous, and many-faceted threat to developing and developed countries that requires multidisciplinary approaches for its effective prevention and combat. Traditional methodologies have made great efforts to understand the phenomenon, but they need the analytical tools to handle the non-trivial structural and dynamical aspects that characterize the modern social, economic, political, and technological systems where corruption presents. In this context, complex systems and network science present themselves as a comprehensive analytical framework for the study of adaptive collective phenomena in complex environments. Therefore, this book gathers several scholars on corruption studies working at the scientific frontier of this phenomenon using the tools of complexity, network science, and social science. The multidisciplinary and global approach of the book is a role model to present theoretical and empirical efforts being performed to curb this problem in which the relevance of evidence-based methods could be the best way to convert the obtained knowledge into public policies against corruption.

Bogotá, DC, Colombia
Mexico, DF, Mexico
May 2021

Oscar M. Granados
José R. Nicolás-Carlock

Acknowledgements

This book has been designed as a creative synthesis of different possibilities to integrate into an interdisciplinary way, complex science, network science, and social sciences to understand corruption. This book has a purpose to opening a long-term perspective that integrates scientists of different disciplines to solve a social problem that affects, systematically, our societies. Since the initial workshop in December 2018, held at Quebec City as a collaboration between the University of Vermont Complex Systems Center and the Sentinel North Program of The Université Laval, we have incurred several debts. We would like to thank the participants in the Complex Networks Winter Workshop 2018, in particular, Juniper Lovato for support, Antoine Allard, Patrick Desrosiers, Samuel Scarpino, and Brooke Foucault Welles for comments and suggestions in our first proposals, as well as the discussions with the integrants of the workshop team, especially power to corrupt. We are also indebted to the participants of Lanet's satellite: Corruption networks, from Data Analysis to Public Policies held at Cartagena, Colombia. We owe special thanks to participants in NetSci 2020 Satellite for their comments. Additionally, it would have been impossible for us to build this book without the comments, advice, and help of numerous colleagues whom we would like to thank here: Laurent Hébert-Dufresne, Marco Tulio Angulo, Brennan Klein, Sebastián Michel-Mata, and Guillermo de Anda Jáuregui. We owe special thanks to our editing team at Springer Nature, Hisako Niko, and Jayanthi Krishnamoorthi. Finally, we should also like to thank our families for their patience during the course of this book.

Contents

Contributors

Tiago Colliri Institute of Mathematics and Computer Science, University of Sao Paulo, Sao Carlos, Brazil

Ágnes Czibik Government Transparency Institute, Budapest, Hungary

Frank Diepenmaat Saxion University of Applied Sciences, Enschede, Netherlands

Mihály Fazekas Central European University, Vienna & Government Transparency Institute, Budapest, Hungary

Carlos Gershenson Instituto de Investigaciones en Matemáticas Aplicadas y Sistemas & Centro de Ciencias de la Complejidad, Universidad Nacional Autónoma de México, Ciudad de México, Mexico;
Lakeside Labs GmbH, Klagenfurt am Wörthersee, Austria

Oscar M. Granados Department of Economics and International Trade, Universidad Jorge Tadeo Lozano, Bogotá, Colombia

Rita Guerrero Plantel Del Valle, Universidad Autónoma de la Ciudad de México, Ciudad de México, Mexico

Gerardo Iñiguez Department of Network and Data Science, Central European University, Vienna, Austria;
Department of Computer Science, Aalto University School of Science, Aalto, Finland;
Centro de Ciencias de la Complejidad, Universidad Nacional Autónoma de México, Ciudad de México, Mexico

Eduardo Islas Subsecretaría de Fiscalización y Combate a la Corrupción, Secretaróa de la Función Pública, Ciudad de México, Mexico

Issa Luna-Pla Institute of Legal Research, National Autonomous University of Mexico, Mexico City, Mexico

José R. Nicolás-Carlock Institute of Legal Research, National Autonomous University of Mexico, Mexico City, Mexico

Carlos Pineda Instituto de Física, Universidad Nacional Autónoma de México, Ciudad de México, Mexico

Omar Pineda Azure Core Security Services, Microsoft, Redmond, WA, USA; Posgrado en Ciencia e Ingeniería de la Computación & Centro de Ciencias de la Complejidad, Universidad Nacional Autónoma de México, Ciudad de México, Mexico

Alfredo Hernandez Sanchez Institut Barcelona d'Estudis Internacionals, Barcelona and Government Transparency Institute, Budapest, Hungary

Willeke Slingerland Saxion University of Applied Sciences, Enschede, Netherlands

Philip C. Solimine Florida State University, Tallahassee, FL, USA

Andrés Vargas Departamento de Matemáticas, Pontificia Universidad Javeriana, Bogotá, Colombia

Johannes Wachs Vienna University of Economics and Business & Complexity Science Hub Vienna, Vienna, Austria

Liang Zhao Faculty of Philosophy, Science, and Letters, University of Sao Paulo, Ribeirao Preto, Brazil

Martin Zumaya Programa Universitario de Estudios sobre Democracia, Justicia y Sociedad & Centro de Ciencias de la Complejidad, Universidad Nacional Autónoma de México, Ciudad de México, Mexico

Chapter 1
Corruption Networks: An Introduction

Oscar M. Granados

> Just as it is impossible not to taste the honey or the poison that
> finds itself at the tip of the tongue, so it is impossible for a
> government servant not to eat up, at least a bit of the king's
> revenue
> —*Kautilya (Arthashastra), c. 350-275 BC*

Abstract To understand corruption, scholars cannot only analyze the details. We need to see the features, the agents, the interactions, the structure, and the dynamics. One way to get the whole picture is to model the corruption processes and systems as a network. This chapter is an introduction both for the corruption networks modeling and the rest of the book. It covers some of the earlier developments from social sciences and complexity that form the foundation for the more specialized topics of the other chapters.

1.1 Corruption as a Social Problem

In a boundary dispute between Damascus and Sidon in Ancient Syria circa AD 30, Agrippa, one of the advisers of Governor Lucius Pomponius Flaccus, accepted a bribe to use his influence to support Damascus interests in the dispute. However, this episode was not an isolated event. Various politicians, advisers, and thinkers in Ancient Egypt, Ancient Greece, Ancient India, the Roman Empire, Imperial China, or later in Renaissance Italy had noticed acts of corruption in some institutions of their governments. Thus, corruption is part of humanity and is not an exclusive problem of modernity or the result of the rise of capitalism. Nor is the result of ethnic origin, educational level, or the economic situation of a population. However, modernity

O. M. Granados (✉)
Department of Economics and International Trade, Universidad Jorge Tadeo Lozano,
Bogotá, Colombia
e-mail: oscarm.granadose@utadeo.edu.co

© The Author(s), under exclusive license to Springer Nature Switzerland AG 2021   1
O. M. Granados and J. R. Nicolás-Carlock (eds.), *Corruption Networks*,
Understanding Complex Systems, https://doi.org/10.1007/978-3-030-81484-7_1

has eventuated in forms of social disorganization has included escalating levels of violence, corruption, and ecological degradation [1].

Corruption has been analyzed from various areas and approaches, mainly from ethics, morals, and law [2, 3], since it is a behavior that goes beyond the simple application of the law and the obedience to law is not taken for granted [4]. Likewise, corruption has been analyzed by anthropology [5], political science [6–11], sociology [1, 12, 13], philosophy [14, 15], and economics [16–18]. The social sciences perspective has established that corruption affects society from different fronts, but always with social costs [19]. The approaches of these sciences have been fundamental to investigate various elements, especially qualitative ones that are necessary to understand the phenomenon of corruption. Additionally, several academics have defined this phenomenon based on the inequality trap and the theory of injustice as part of its advance [20–22]. From economics, the validation of corruption impact has also been analyzed from inequality and poverty perspectives. Additionally, corruption affects economic freedom [23, 24], economic growth [25, 26], investment processes [27, 28], and economic development [29–31]. These arguments have defined that corruption is deeper in countries with economic difficulties, which in several cases is true [32–34]. However, empirical evidence had shown corruption in countries with high levels of foreign investment, economic growth, and economic development, as well as income equality [35]. Corruption is not a phenomenon exclusive to developing countries or those mired in poverty, corruption is more visible in those countries because it is not possible to understand how in these countries with problems in the provision of utilities, education, and health, a public officer enjoy privileges that his salary would not allow. Likewise, although some authors mention that countries with high economic freedom are less prone to corruption, it does not mean that they are exempt from it. Furthermore, the corruption from the organizations is a relevant element in corruption processes [36, 37].

It seems that our society has become accustomed to a corrupt framework since corruption acts are more frequent at various levels. From billionaire public contracts to everyday life where small corruption processes arise in exchange for paying a few coins. Those macro and micro corruption processes have accompanied our society with different manifestations that lead to other crimes such as discrimination, influence peddling, fraud, tax evasion, or money laundering, to list a few. The corruption framework has gone so far that, in some countries, the only way to win a contract or obtain a public document is by somehow paying a benefit to an official. Every corruption act needs the procedures to achieve its final purpose: economic benefit.

Does a great concern arise on why people do not stop at any conflict of interest? Could the path to corrupt actions make more difficult? Interest conflicts are not corrupt's concern, and they accelerated the intention to obtain benefits from them without any qualms. Namely, corruption has progressed smoothly to convert on a systemic phenomenon in which particular interests outweigh collective interests. Corruption arises as part of the human condition and is connected with the capacities to do something, as Arendt [38] argues when referring to the human capacity for thought. However, with its advances, corruption is closer to being an attribute of the human-animal, i.e., corruption is a part of human essence and daily life. This conclusion may be simple speculation, but after more than 2,000 years of documented corrupt acts, corruption is part of humanity to the full extent of the word.

Social sciences have identified reasons and consequences of corrupt actions with more than 10.000 articles and technical documents. However, it is necessary to identify the dynamics of corruption since they go beyond the interaction of a couple of persons or the dominant position that induces this type of action. Likewise, it is necessary to identify the evolution of corruption since it is an action that is increasing and is repeated generation after generation, expanding more and more at different levels of public and private contexts. However, corruption tries to go unnoticed, unidentified, to hide in changing scenarios and different temporalities. In other words, corruption is a complex system with open and adaptive features, which for its analysis requires a group of tools in permanent transformation.

1.2   Corruption, Complexity, and Networks

Several authors of complexity science have approached the analysis of social problems. Corruption emerges as part of a human social environment, and it requires the complexity profile proposed by Bar-Yam [39] as a mathematical tool to characterizing the collective behavior of a system. Additionally, since corruption tends towards a progressive appearance of collective behaviors of ever-larger groups of people, it becomes a system that grows in complexity but also in adaptation, i.e., adaptive complex systems [40–42]. Thus, unlike Agrippa's case in Ancient Syria, where his corrupt actions involved only him, in the 21st century, corruption integrates increasingly large groups of people, possibly due to the actions that must be carried out to obtain the benefit.

The discouraging reports of Transparency International [43, 44] and international organizations as the United Nations, World Bank, and International Monetary Fund, as well as the corruption problems in different countries like Brazil, China, Colombia, Haiti, India, Italy, Mexico, Nigeria, Philippines, Venezuela, and a long list of countries, have increased the interest of scholars of complexity. From approaches with applied perspective or pure theoretical perspectives [45–54], the corruption studies have won new investigations in the last years that identify corruption as a complex system where statistical mechanics, nonlinear dynamics, complex networks, artificial intelligence, and other methodologies consolidate new approaches to social problems like sociophysics [55], but also a combination of methods.

Complex systems consist of diverse elements that self-organize, driven by their random interactions, into ordered systems that exhibit feedbacks and nonlinearities. Complex systems are exposed to perturbations but also some features that emerge from the interaction with the context. Thus, we need to understand corruption networks as a complex networks [56–60]. First, corruption networks are involved with the randomness of interactions that create an evolutionary process. Second, corruption networks evolve when they do not have limits, generating cascading processes that affect public and private resources. Third, the diffusion process of corruption can accelerate the collapse of institutions and organizations. Fourth, the policies and tools to fight corruption are urgent because if corruption networks elevated to greater diversity (more agents and interactions), the higher is its survival probability. Fifth, if we understand the dynamics of corruption networks, the tools, and hence the policies, should be to restrict the dynamics and openness of the corrupt system. Sixth, if we identify the feedback process of corruption networks, the tools could limit it to reduce the consolidation of the corruption complex system.

I have given an overview of the corruption networks. Furthermore, I have explained the challenges in extending the corruption analysis to a multidisciplinary framework because that phenomenon is not an issue of one discipline, the corruption is a human problem that needs an elite group of diverse methods that create tools to fight it. The challenges to study corruption are intellectually fascinating, but also they have social retribution since useful tools are waiting to apply in different places of the world. However, many challenges remain. The studies of corruption and related crimes as tax evasion and money laundering need a dynamic perspective of methods to identify the evolution of these phenomena. Network science, complex systems, dynamic systems, statistical mechanics, topology, geometry, artificial intelligence are several methods to fight corruption evolution, a social problem that social sciences have fought for years. Those methods are a methodological complement to social science to reach a phenomenon that has taken us a great advantage. Future works range to fight corruption effectively as how to use geometry to identify corruption cycles and how to make simulations of corrupt systems to create public policies. Hence, this book is the first result of a group of scholars that needs other scholars that join this enterprise to solve a problem that could be as big as climate change.

1.3 The Structure of the Book

Writing an analysis of corruption represents a challenge for any scientist. How can corruption networks, which make up the subject of this book, be separated from the national environment when corruption advance in many places worldwide, and the suspicious agents used different instruments to hide their actions? What is the exact purpose of this book? This book aims at presenting an overview of the state-of-the-art in corruption networks. Its chapters are contributed by researchers and research teams from a variety of backgrounds, disciplines, and approaches to corruption. Our target has been to cover the emerging field of corruption networks both in breadth and in-depth, and because of this, some chapters are reviews on relevant topics such as social capital and control theory in corruption networks, whereas others provide detailed accounts of investigations building on corruption processes using network science framework, from a local or specific topic or an international network to a new perspective about corruption analysis.

The following chapters focus on different methods for characterizing corruption networks. In the chapter social capital, corrupt networks, and network corruption, Willeke Slingerland presents a different outlook on the link between networks and corruption. She develops the concept of network corruption, which is the phenomenon of collective acting by networks, which results in corruption even if the individual acting itself is not necessarily corrupt. The concept considers corruption as a social process within networks. This is followed by a chapter that focuses on one of the relevant features of corruption networks, especially those related to control. In the chapter Network controllability metrics for corruption research, Philip Solimine provides us with an account of how metrics of network control theory facilitate understanding the mechanisms by which corrupting actors can perturb nodes' behavior at certain points within a social system, especially the political corruption in a legislative social network. The chapter analyzes as corrupting actors harness perturbations to drive the network to a desirable state where they can amplify the effects of corruption using a social influence network or hierarchy.

Then, in the chapter Predicting corruption convictions among Brazilian representatives through a voting-history based network, Tiago Colliri and Liang Zhao address the voting data concerning almost 30 years of legislative work from Brazilian representatives, focusing on identifying the formation of corrupt neighborhoods in the resulting congresspeople network through a predictive model for assessing the chances of a representative for being convicted of corruption or other financial crimes in the future, solely based on how similar are his past votes and the voting record of already convicted politicians. In the chapter Networked Corruption Risks in European Defense Procurement, Agnes Czibik, Mihály Fazekas, Alfredo Hernandez Sanchez, and Johannes Wachs study Defense procurement to develop an objective corruption risk indicator. They identify that risk indicator is higher for military contracts than for contracts in general, and the corruption risk is significantly higher in the periphery, while in others, it is higher in the center. This chapter connects the corruption networks and economic crimes based on corruption. In the chapter Identifying tax evasion in Mexico with tools from network science and machine learning, Martin Zumaya et al., analyze with those methods more than 80 million contributors and almost 7 billion monthly aggregations of invoices among contributors to identify tax evaders. They build temporal networks where nodes are contributors, and directed links are invoices produced in a given time slice and show that their interaction patterns differ from those of the majority of contributors.

This chapter on tax evasion connects with another economic crime: money laundering. From a legal perspective, Frank Diepenmaat describes in this chapter the international legal framework for the repression and the prevention of money laundering from an existing network in place to fight this crime and initiatives to create new networks with better cooperation between government institutions and private organizations. In a second perspective, Oscar Granados y Andres Vargas analyze the large-scale structure of global financial networks and focus on particular aspects of their characteristics when suspicious activities of tax fraud, corruption, and money laundering could be identified. They reveal that suspicious activities run in small groups, and they emerge around communities of financial intermediaries, non-financial intermediaries, and offshore entities. Finally, José Nicolás-Carlock and Issa Luna present some problems in corruption analysis that need to analyze them as the complexity of the corruption phenomenon itself and its context, the complexity of the analytical description, and the complexity of the perspectives that different disciplines bring to the table. They argue that the interdisciplinary framework of complex systems and network science represents a promising analytical approach to move a new inter-disciplinary framework for corruption studies: corruptomics. This final chapter is not an epilogue. It is the first input to create an open system where diverse scholars, using methodologies from physics, mathematics, computer science, and social science, are interested in developing tools to help to reduce the corruption of our corporations, institutions, towns, cities, regions, countries, but principally to reduce it on the next generations.

References

1. Clammer J (2012) Corruption, development, chaos and social disorganisation: sociological reflections on corruption and its social basis. ANU Press, pp 113–132. https://doi.org/10.2307/j.ctt24hbwc.11
2. Rothstein B (2018) Fighting systemic corruption: The indirect strategy. Daedalus 147(3):35–49 https://doi.org/10.2307/48563079
3. Rose-Ackerman S (2018) Corruption & purity. Daedalus 147(3):98–110. https://doi.org/10.2307/48563083
4. Becker GS (1968) Crime and punishment: An economic approach. J Polit Econ 76(2):169–217. https://doi.org/10.2307/1830482
5. Robertson AF (2006) Misunderstanding corruption. Anthropology Today 22(2):8–11. https://doi.org/10.2307/3695271
6. Nye JS (1967) Corruption and political development: A cost-benefit analysis. American Political Science Review 61(2):417–427. https://doi.org/10.2307/1953254
7. Nelken D, Levi M (1996) The corruption of politics and the politics of corruption: An overview. J Law Soc 23(1):1–17. https://doi.org/10.2307/1410464
8. Cartier-Bresson J (1997) Corruption networks, transaction security and illegal social exchange. Political Studies 45(3):463–476. https://doi.org/10.1111/1467-9248.0009
9. Andersen CJ, Tverdova YV (2003) Corruption, political allegiances, and attitudes toward government in contemporary democracies. American Journal of Political Science 47(1):91–109. https://doi.org/10.1111/1540-5907.00007
10. Heywood PM (2017) Rethinking corruption: Hocus-pocus, locus and focus. The Slavonic and East European Review 95(1):21–48 https://doi.org/10.5699/slaveasteurorev2.95.1.0021
11. Gerring J, Thacker SC (2004) Political institutions and corruption: The role of unitarism and parliamentarism. British Journal of Political Science 34(2):295–330. https://doi.org/10.2307/4092365
12. Warburton J (2013) Corruption and anti-corruption. In: Chapter Corruption as a social process: from dyads to networks. ANU Press, Canberra, pp 221–237. https://doi.org/10.22459/CAC.03.2013.13
13. Kim HJ, Sharman JC (2014) Accounts and accountability: corruption, human rights, and individual accountability norms. Int Organ 68(2):417–448. https://doi.org/10.2307/43283272
14. Saxonhouse AW (2012) To corrupt: the ambiguity of the language of corruption in ancient Athens. ANU Press, pp 37–52. http://www.jstor.org/stable/j.ctt24hbwc.7
15. Sparling R (2016) Montesquieu on corruption: civic purity in a post-republican world. University of Toronto Press, pp 157–184. https://doi.org/10.3138/j.ctt1kk65xt.12
16. Banfield EC (1975) Corruption as a feature of governmental organization. The Journal of Law & Economics 18(3):587–605. https://doi.org/10.2307/725047
17. Davoodi H, Tanzi V (1997) Corruption, public investment, and growth. IMF Work Pap 97:11. https://doi.org/10.5089/9781451929515.001

18. Tanzi V (1998) Corruption around the world: causes, consequences, scope, and cures. IMF Work Pap 45(4)
19. Spencer H (1843) The proper sphere of government. The Nonconformist
20. Rothstein B, Uslaner EM (2005) All for all: Equality, corruption, and social trust. World Politics 58(1):41–72. https://doi.org/10.1353/wp.2006.0022
21. You J-S, Sanjeev K (2005) A comparative study of inequality and corruption. Am Sociol Rev 70(1):136–157. https://doi.org/10.1177/000312240507000107
22. Ariely G, Uslaner EM (2017) Corruption, fairness, and inequality. International Political Science Review / Revue internationale de science politique 38(3):349–362. https://doi.org/10.2307/44632346
23. Martin Paldam. The cross-country pattern of corruption: economics, culture and the seesaw dynamics. European Journal of Political Economy, 18(2), 215–240, 2002. https://doi.org/10.1016/S0176-2680(02)00078-2
24. Malanski LK, Póvoa ACS (2021) Economic growth and corruption in emerging markets: does economic freedom matter? Int Econ 166:58–70. https://doi.org/10.1016/j.inteco.2021.02.001
25. P. Mauro. Corruption and growth. Quarterly Journal of Economics, 110(3), 681–712, 1995. https://doi.org/10.2307/2946696
26. P.H. Mo. Corruption and economic growth. Journal of Comparative Economics, 29(1), 66–79, 2001. https://doi.org/10.1006/jcec.2000.1703
27. Ades A, Di Tella R (1999) Rents, competition, and corruption. American Economic Review 89(4):982–993. https://doi.org/10.1257/aer.89.4.982
28. M. Habib and L. Zurawicki. Corruption and foreign direct investment. Journal of International Business Studies, 33(2), 291–307, 2002. https://doi.org/10.1057/palgrave.jibs.8491017
29. N.H. Leff. Economic development through bureaucratic corruption. American Behavioral Scientist, 8(3), 8–14, 1964. https://doi.org/10.1177/000276426400800303
30. Bardhan P (1997) Corruption and development: A review of issues. Journal of Economic Literature 35(3):1320–1346
31. Treisman D (2000) The causes of corruption: A cross-national study. J Public Econ 76(3):399–457. https://doi.org/10.1016/S0047-2727(99)00092-4
32. A. Shleifer and R.W. Vishny. Corruption. Quarterly Journal of Economics, 108(3), 599–617, 1993. https://doi.org/10.2307/2118402
33. Aidt TS (2003) Economic analysis of corruption: A survey. Economic Journal 113(491):F632–F652. https://doi.org/10.1046/j.0013-0133.2003.00171.x
34. Svensson J (2005) Eight questions about corruption. Journal of Economic Perspectives 19(3):19–42. https://doi.org/10.1257/089533005774357860
35. Di Vita G (2021) Political corruption and legislative complexity: Two sides of same coin? Struct Chang Econ Dyn 57:136–147. https://doi.org/10.1016/j.strueco.2021.03.004
36. Djankov S, Porta RL, Lopez-de Silanes F, Shleifer A (2002) The regulation of entry. Quart J Econ 117(1):1–37. https://doi.org/10.1162/003355302753399436
37. Ashforth BE, Anand V (2003) The normalization of corruption in organizations. Research in Organizational Behavior 25:1–52. https://doi.org/10.1016/S0191-3085(03)25001-2
38. Arendt H (1958) The Human Condition. The University of Chicago Press, Chicago
39. Bar-Yam Y (2002) Encyclopedia of Life Support Systems, chapter Complexity rising: From human beings to human civilization, a complexity profile. EOLSS Publishers, Oxford
40. Gell-Mann M (1995) Quark and the Jaguar. Adventures in the Simple and the Complex. Henry Holt & Company, New York
41. Bar-Yam Y (1997) Dynamics of Complex Systems. Perseus Books, Reading, MA
42. Miller JH, Page SE (2009) Complex adaptive systems: an introduction to computational models of social life, vol 17. Princeton University Press, Princeton
43. Pring C (2017) People and corruption: citizens voices from around the world. Work Pap, Transp Int
44. Transparency International (2021) The corruption perceptions index 2020. Technical report, Transparency international

45. Raúl C (1999) Large-scale corruption: definition, causes, and cures. Syst Practic Action Res 12(4):335–353. https://doi.org/10.1023/A:1022492112414
46. Ribeiro HV, Alves LGA, Martins AF, Lenzi EK, Perc M (2018) The dynamical structure of political corruption networks. J Complex Netw 6(6):989–1003. https://doi.org/10.1093/comnet/cny002
47. Johannes W, Taha Y, Balázs L, János K (2019) Social capital predicts corruption risk in towns. R Soc Open Sci 6(4):182103. https://doi.org/10.1098/rsos.182103
48. Johannes Wachs and János Kertész. A network approach to cartel detection in public auction markets. Scientific reports, 9(1):1–10, 2019
49. Fazekas M, Wachs J (2020) Corruption and the network structure of public contracting markets across government change. Politics and Governance 8(2):153–166
50. Wachs J, Fazekas M, Kertész J Corruption risk in contracting markets: a network science perspective. Int J Data Sci Anal. https://doi.org/10.1007/s41060-019-00204-1
51. Kertész J, Wachs J (2020) Complexity science approach to economic crime. Nat Rev Phys. https://doi.org/10.1038/s42254-020-0238-9
52. Luna-Pla I, Nicolás-Carlock JR (2020) Corruption and complexity: a scientific framework for the analysis of corruption networks. Applied Network Science 5(1):13. https://doi.org/10.1007/s41109-020-00258-2
53. Lu D, Bauza F, Soriano-Paños D, Gómez-Gardeñes J, Floría LM (2020) Norm violation versus punishment risk in a social model of corruption. Phys Rev E 101:022306. https://doi.org/10.1103/PhysRevE.101.022306
54. Garcia-Bedoya O, Granados O, Burgos JC (2020) AI against money laundering networks: the Colombian case. J Money Laund Control 24(1):49–62. https://doi.org/10.1108/JMLC-04-2020-0033 https://
55. Galam S (2012) Sociophysics. A Physicist's Modeling of Psycho-political Phenomena. Springer-Verlag, New York
56. Watts Duncan J, Strogatz Steven H (Jun 1998) Collective dynamics of 'small-world' networks. Nature 393(6684):440–442. https://doi.org/10.1038/30918
57. Boccaletti S, Latora V, Moreno Y, Chavez M, Hwang D-U (2006) Complex networks: Structure and dynamics. Phys Rep 424(4):175–308. https://doi.org/10.1016/j.physrep.2005.10.009
58. Alex Arenas, Albert Díaz-Guilera, Jurgen Kurths, Yamir Moreno, and Changsong Zhou. Synchronization in complex networks. Physics Reports, 469(3), 93–153, 2008. https://doi.org/10.1016/j.physrep.2008.09.002
59. Estrada E (2011) The Structure of Complex Networks: Theory and Applications. Oxford University Press, Oxford
60. Chen D, Lü L, Shang M-S, Zhang Y-C, Zhou T (2012) Identifying influential nodes in complex networks. Physica A 391(4):1777–1787. https://doi.org/10.1016/j.physa.2011.09.017

Chapter 2
Social Capital, Corrupt Networks, and Network Corruption

Willeke Slingerland

Abstract The network literature available presents valuable theories on corruption networks. These networks are goal-directed criminal networks, established to undertake criminal activities. This chapter presents a different outlook on the link between networks and corruption. It presents the concept of 'network corruption' which is the phenomenon of collective acting by networks which results in corruption even if the individual acting itself is not necessarily corrupt. This chapter is partly based on earlier cases studies which were investigated as corruption cases by the public prosecution services. People make group-serving or 'sociocentric' attributions to boost the performance of groups to which they belong, and this can turn into network corruption. The concept of network corruption considers corruption as a social process within networks. Its introduction implies the need to understand corruption as collective acting and thereby underlining the importance to incorporate network awareness and network responsibility in corruption literature and in practical policy making. Network corruption as a complex system provides insights into norm development in emerging networks which all of us need to keep an eye on when we involve and participate in networks, whether we participate as politician, civil servant, law enforcement official, journalist, entrepreneur, employee or citizen.

2.1 Introduction

Corruption undermines human development. It diverts public resources away from the provision of essential services. It increases inequality and hinders economic development by distorting markets for goods and services. It undermines rule of law and democracy, foremost because it destroys public trust in governments and leaders [2]. In many countries across the world corruption cases make the headlines. Expos-

Parts of this chapter have been published in "Network Corruption: When Social Capital Becomes Corrupted. Its meaning and significance in corruption and network theory and the consequences for (EU) policy and law" [1].

W. Slingerland (✉)
Saxion University of Applied Sciences, Enschede, Netherlands
e-mail: w.slingerland@saxion.nl

© The Author(s), under exclusive license to Springer Nature Switzerland AG 2021
O. M. Granados and J. R. Nicolás-Carlock (eds.), *Corruption Networks*,
Understanding Complex Systems, https://doi.org/10.1007/978-3-030-81484-7_2

ing corruption and holding the corrupt to account can only happen if we understand the way corruption works and the systems that enable it [3]. The current literature and research on corruption in network-like structures mostly focusses on systemic corruption in the sense that most network members are involved in bribe paying of any sort. Ashforth and Anand [4] present a model to examine collective corruption. Although they still build their model on the idea of corrupt acts, their research is a step in the direction of understanding corruption to be embedded in collective structures and processes. They argue that there are three reinforcing and reciprocally inter-dependent processes (institutionalization, rationalization and socialization), which conspire to normalize and perpetuate corrupt practices so that the system beats the individual. Warren [5] points to the fact that in any democracy second-order norms of process can evolve that can quite properly bring new meanings of corruption in their wake. Zimelis [6] and Vergara [7] both argue that the corruption studies suffer from the bias of focusing on systemic corruption in developing countries and nation-states while an integrated approach to studying systemic corruption in western countries is needed. Vergara argues that the systemic corruption takes on the form of an oligarchic democracy, a non-representative liberal government.

2.2   Corruption

2.2.1   Existing Typology of Corruption

Before continuing with the exploration between corruption and networks, it is impor-tant to understand how scientists have explored the corruption concept. The corrup-tion literature is characterized by the following polytomy: corrupt individuals ('bad apples') versus corrupt organizations ('bad barrels') versus corrupt systems ('bad barrel makers'). Corruption cases all have in common that somebody with power and influence knowingly uses his influence to e.g. assign a public contract, provide a permit license, refrain from carrying out safety measures, influence policies or laws and in return receives a favor given by a third party, who acted on behalf of themselves or others. Notwithstanding the possibility that extortion or other means of pressure were used to get these persons to do so, this way of acting was a conscious and rational choice of the individual; it is the individual who is corrupt. The 'bad apple' perspective blames the immoral individual whose personal characteristics cause the unethical behavior. As such, corruption is foremost a matter of rational choice or a calculated decision of an individual who instigates or accedes to an attempted corrupt transaction [8, 9].

Another approach is to consider corruption to be something at the level of the organization. The 'bad barrels' perspective links unethical and immoral organiza-tional culture to corrupt practices. Pinto et al. [10] distinguish two distinct corruption phenomena at the level of the organization. The first appearance of corruption con-cerns the cases in which a significant proportion of an organization's staff act in a corrupt manner but primarily for their own personal benefit (an organization of corrupt individuals). The second appearance concern the cases in which a group collectively acts in a corrupt manner for the benefit of the organization (a corrupt organization). Here, corruption can be seen as a social process, within the context of the organization, allowing the group to be organized in such a way that it commits corruption on behalf of the organization [9, 10].

Besides seeing corruption as a matter of individual choice or seeing corruption as related to the organizational context, corruption is also regarded as being understood from a system or 'bad barrel maker' perspective. Individuals and organizations are embedded in a larger societal system which influences their behavior. Johnston [11] is at the forefront in the academic debate when stressing the importance of understanding the system level of corruption. He understands corruption from the setting in which it evolves, both in terms of a country's level of economic development and its level of democracy. This system level of corruption is acknowledged by scholars, policy makers and NGOs such as Transparency International. Klitgaard [8] and Nielsen [12] have tried to capture the elements of corrupt systems and introduced strategies to fix the systems that breed corruption. This system approach is reflected in Transparency International's model of National Integrity System (NIS) and the model Local Integrity System (LIS) by Huberts and Six [13]). Both models understand corruption to emerge if there are weakness within societal foundations (economic, social, cultural) and both hard controls (e.g. strong and independent institutions, rules an checks and balances) and soft controls (e.g. public education, leadership) are not aligned.

2.2.2   Adding a New Classification of Corruption to the Typology

The exploration of the body of knowledge on corruption shows that corruption is a layered concept which leads to the conclusion that there is no definition of corruption which is universally agreed upon. Corruption literature does present a system approach of corruption, mostly by looking at how individual corrupt acts are in fact caused by the collective behavior of groups and organizations, here corruption becomes the norm and many are committing corruption. The other systems approach to corruption is understanding its emergence by pointing at the specific weaknesses within a given societal system. Both approaches imply a focus of corruption as an individual act in a certain context. Having assessed the most used corruption definitions and typologies, I have come to define corruption as "allowing improper interests to influence decision-making at the expense of the general interest". This definition not only meets the idea of corruption as an individual act, it also meets the idea of corruption as collective acting even when individuals do not act corrupt.

The cases which were part of the work this article is based on were investigated as corruption cases by several national public prosecution services. All three cases involved various other misconducts such as favoritism, abuse of power and conflicts of interests. Although large-scale corruption was suspected, the limited scope of the bribery provisions required evidence of each individual's abuse of power in return for a bribe or it had to be proven that an individual paid a bribe. In all three case studies such form of corruption could only be proven for a few individuals or not proven at all. This is not to say that there was no serious corruption, the corruption emerged in a more subtle way. It went beyond the level of individual acting and immediate and obvious influencing. Before going into detail, the following two visualizations are presented to illustrate the form of corruption which will be introduced as network corruption.

This network perspective is visualized in image 2 where a random local politician's personal and professional connections are all interwoven. The average corruption risk analysis looks at possible stakeholders and vulnerable bilateral relations as can be seen in image 1. However, when we incorporate networks into such a risk analysis other 'red flags' become visible. The center network in red is the small network of the politician and his family. The pink network is composed of the connections which are related to these family members (school, leisure and friends). The purple network is a representation of networks which developed during previous professions and studies. The blue network is the local governmental organization. The orange network represents the connections within the network of the political party to which the politician is affiliated. The green network consist of connections related to the portfolio of the politician. The politician himself and many other key figures visualized in this image all have power and influence which coincide with their formal positions and additional functions. Besides the formalized networks which coincide with these positions, they are part of more informal networks. Many of these networks overlap and directly create a risk of conflict of interest. These networks allow improper interests to influence decision-making at the expense of the general interest. This mechanism is seen in many of the official court and legal documents of the case studies. This is where current network and corruption literature fall short and the day to day reality of corruption prosecution is frustrated because here corruption is a collective network outcome developed over time without the easy to point at 'quid pro quo'. To be able to understand this corruption emergence, we need to use network theory and its explanations of mechanisms within social networks. Before presenting these mechanisms, a short distinction between several types of networks is presented.

2.3   Social Networks

Figure 2.1a and b have visualized the difference between the perspective on ones' bilateral or multilateral relations and what this could imply for corruption research. These formal and informal networks are all social networks which form the structure in which we as human beings connect with others. A social network is a set of actors and the set of ties representing some relationship or lack of relationship between the

a

b

Fig. 2.1 This visualization is based on an example of a local politician and the formal and informal networks a politician is a member of. Such a network image can be made for any individual and is not limited to the political or local context. **a.** Visualization of a random local politician and his (bilateral) personal connections. **b.** Visualization 2 of a random local politician and how his personal connections are all interwoven

actors [14]. It is an arrangement of people crossed at regular intervals by other people, all of whom are cultivating mutually beneficial, give-and- take, win-win relationships [15]. So each social network is composed of actors and relations. Actors can be individual natural persons or collectives. Relations can best be described as the specific kind of connection or tie between these actors. Besides these basic characteristics of networks, we need to make a distinction between formal and informal networks and between goal-directed and emergent networks and take a closer look at the general characteristics of social networks. These will allow a further understanding of the link between networks and when they create social capital or corruption.

2.3.1 Formal and Informal Networks

Social networks can take on different forms. Social networks are not mere contacts, they imply mutual obligations [16]. An individual's network entails everyone from family members, friends and acquaintances to professional contacts. The visualizations above include a few well-known and obvious networks of a local politician. We can distinguish formal networks such as student associations, business clubs or political parties. These networks have a formal structure and formal hierarchy of their own. One facet of social connectedness is the official membership of such a formal organization, the other facet includes memberships of informal social networks. A social network distinguishes itself from official organizations because it is not formally founded but instead has an informal horizontal structure in which these beneficial relationships emerge. Putnam [16] distinguishes formal 'card carrying' membership from actual informal involvement in community activities, whereby the latter has more effect in terms of value. Networking requires individuals to stay in touch with their contacts formally or informally to know what they are up to and to stay informed about their needs. Actions of individuals and groups can be facilitated by their memberships in social networks, foremost because of their direct and indirect links to others in these networks. Social networks are not abstract concepts, they are the informal structures which we all find ourselves in while working or meeting friends. They do have effects which are real and noticeable. In this article we focus on the corruption effect of networks. Besides the distinction between formal and informal networks, we also need to consider the difference between goal-directed and emergent networks, in order to understand network goals outcomes.

2.3.2 Goal-Directed Networks and Emergent Networks

In order to understand the link between networks and corruption we also need to understand the coming into existence of networks. The literature on whole networks or (inter)organizational networks provides important insights into the way networks develop, how they are functioning and how social networks are embedded in larger 'whole networks'. Whole networks are defined as three or more autonomous organizations collectively working together to achieve not only their individual organizational goals, but also a common network goal. These whole networks are complex systems and can either be goal-directed or emergent. Whole networks are a group of three or more organizations connected in ways that facilitate achievement of a common goal [17]. Such networks are often formally established and governed, and goal-directed, rather than occurring serendipitously. Relationships among network members are primarily non-hierarchical and participants often have substantial operating autonomy. Goal-directed networks are maintained in a more formal way, such as by means of contracts and rules [17]. Their relationships are maintained informally through the norms of reciprocity and trust. The multilateral relations define a whole network and are essential for achieving collective outcomes. Such networks are critical for resolving many of the problems and policy issues that office holders are confronted with. In many countries network involvement is welcomed because it is seen as an effective way to realize private sector consultation and participation [18].

Another category of networks is that of emergent networks. These networks develop organically and over time. Factors such as friendship, trust, or the need to acquire legitimacy or power are the basis of successful relationships [19]. Vilkas [20] revealed that emergent networks enable collective micro-processes such as negotiation, adaptation and integration thereby influencing formal structures and processes. As such, the impact of emergent networks should not be overlooked.

2.3.3 General Characteristics of Networks

In the following section, we will introduce key characteristics from social network science that will play a role in our analyses of networks and corruption. Although there are variations among networks, there are some features which all social networks have in common. Social networks have some distinguishing features: 1. The main characteristic of a social network is the absence of a formal structure and the presence of an informal structure. Networks are flat and horizontal, whereas formal organizations are governed by means of a certain hierarchy. Small groups add an important element to the way in which modern life is organized. It is a change from formal organizations in the direction of interpersonal unorganized or loosely organized connections [16].

2. A second characteristic is that a network does not have a formal leader or director but so-called centers of influence can be distinguished. This center of influence is the actor which has been part of a network for a long time or who has a high-profile position [15]. Although the actor may not be able to offer someone immediate advantages, he or she may be able to connect someone with others who can, while this connection might be useful in the future. The foundation for networking is based on the old expression "it's not what you know, but who you know" [15]. Granovetter [21] did research into the strength of interpersonal ties and as an example showed

that although person A has a close connection with B and A also is closely connected to C, even though B and C have no relationship, common ties to A will at some point bring about a certain interaction. There is a likelihood that friendly feelings emerge once they met because of their expected similarity to A. A in this example can be seen as a center of influence in such a network.

3. A third characteristic is that social networks are dynamic, responsive and adaptive. Formal organizations are rather static and bureaucratic. Social networks are systems which consist of nodes (individual actors, people, or things within the network) and the ties, links or connections (relationships) between them. These nodes can be tangibles such as individual persons, regulations, institutions, permits or money, and intangibles, for instance, success, pride or feelings of loyalty and fear. The interconnections are the relationships which hold these elements together. Flows of information are the most important and common form of interconnections and become signals that trigger decision or action points in a system [22]. If a consistent behavioral pattern over a long period of time can be seen, this means that the feedback loop mechanisms are working.

4. The fourth characteristic of the social network is that it has a purpose. The goal or purpose of any network can best be characterized by looking at the conduct of the network members. This conduct reflects the collective goal and it can be different from goals formalized on paper. Brass et al. [23] note that those who interact within networks become more similar. As such, the common attitude of connected individuals is not only proof of the existence of a network; it is also evidence of the network's purpose. Goal-directed networks have been established with a goal in mind, while the emergent networks strive for a certain goal but this process is the outcome of an organically grown network. Social networks create an opportunity that may facilitate access to a variety of resources (finances, information and political support) which ensures one has adequate means to exist and be successful in a competitive society.

5. A fifth important characteristic of a network is that of 'actor similarity' [16, 23]. Similar people tend to interact with each other, or 'likes attract likes'. Similarity is thought to ease communication, increase the predictability of behavior, and foster trust and reciprocity. Individuals identify themselves with the network to which they are a member. Turner et al. [24] introduced this idea of 'self-categorization'. When people 'self-categorize', they are motivated by social group goals rather than individual goals, make the group's characteristics their own personal characteristics, and thus incorporate social group identities into their personal identities [25]. In social psychology, social identity theory is based on the idea that membership in social groups is an important determinant of individual behavior [25]. Social networks coordinate an individual's behavior and ethical decision-making in a certain direction. Social identity theory explains how a person has a social group identity in terms of 'identifying with' the collective [25]. As soon as a person identifies with others, their individual identity is re-framed in terms of others' identities. This implies that individuals contribute to the identity of the social group, which in turn influences their own identity. Ashforth and Anand [4] refer to a similar process being that of socialization. This process involves imparting to newcomers the values, beliefs,

norms and conduct, that they will need to fulfil their roles and functions effectively within the group context. As individuals seek to make sense of reality, they compare their own perceptions with those of others [23]. Ashforth and Anand [4] use the metaphor of a 'social cocoon' or 'microcosm' where newcomers of a group are encouraged to affiliate and bond with older members, fostering desires to identify with, emulate and please them. As such, they not only identify with the other group members and the collective they form together, but also identify with their own role in the collective [4].

The psychological and sociological mechanism of wanting to belong to a group and being less critical towards the conduct of in-group members has had implications in other disciplines. There is a legal concept named 'willful blindness', which refers to a situation in which a person intentionally fails to be informed about matters that would make one criminally liable [26]. So there is a possibility to know something and a responsibility to be informed, but it is avoided. Willful blindness originates from the fact that human beings have a tendency to build relationships that reaffirm their values, make them feel comfortable and blind them to alternatives. This form of negligence within networks has led to environmental disasters and the financial crisis [26]. This mechanism is likely to play a role in social networks examined by this study.

6. The sixth characteristic of a network is that it is not simply the sum of its members. Major societal successes and concerns are embedded in social systems. If a phenomenon is caused by a collective, but cannot be reduced to individual acts and members, it should be addressed as an independent actor.

2.4 Social Capital

The previous section dealt with the differences between formal and informal and goal-directed and emergent networks. These various networks all form a type of social capital that encompasses the connections that people have with other professionals, family, friends and acquaintances and which brings with it potential advantages granted by members of the network because they have power and influence. The relationships are based on the principle that people have a mutual concern for each other's welfare and well-being. This "the need to belong" is a human desire for close relationships and a strong human motivation [27]. This desire for social connectedness is analogous to our needs for food and water [28]. What differentiates social capital from other forms of capital is that this capital is not located in the actors but in the relations they have with others [29]. Social capital, as defined by Putnam, "refers to networks and the norms of reciprocity and trust that arise from them" [29]. Fukuyama [30] states that "social capital is an instantiated informal norm that promotes co-operation between two or more individuals." The OECD [31] defines social capital as "networks together with shared norms, values and understandings that facilitate co-operation within or among groups". Most scholars consider the source of social capital to be located in the structure of the network [16, 29].

The reward of the social interactions is intrinsic. There is considerable consensus among scholars that the social interaction itself is the reward [32]. Network members might get economic advantages through the social networks, but this is not the initial motive for forming a network. According to Brass et al. [23] social capital is "getting ahead to be a matter of who you know, not what you know". The people who do better in society are somehow better connected [33]. Certain people or certain groups are connected to certain others, trusting certain others, obligated to support certain others, dependent on exchange with certain others. Holding a specific position in the structure of these exchanges can be an asset in its own right [33].

The social network is seen as an important social organization or structure in our society. Scholars such as Granovetter [21] and Vilkas [20] have pointed to the failure to recognize the importance of these networks of personal relations in the economic system, while Fukuyama states that a loosely organized civil society serves to counter the power of the state and to protect individuals from the state's power. If such social capital is absent, there is a risk of centralized tyranny and poor governance which can be the seed for corruption [30]. Warren [5] connects social capital to today's democracies by stressing that democracies work when their people have capacities to associate for collective purposes. Human beings' ability to organize themselves in networks is essential to ensure interest representation in a well-balanced society, economy and political environment. These networks are crucial in any democracy, whether it is an established and resilient democracy, a challenged democracy or a developing democracy. So far, the focus has been on the good side of social networks. However, the case studies this article is based on also indicate that there are risks involved in networking. Many corruption cases are brought in connection with networks and connections. Therefore the question remains what other effects or outcome social networking can bring about.

2.5 Deterioration of Social Capital

2.5.1 Legal Versus Illegal Networks

The previous paragraphs dealt with the positive side of social networks: human beings' ability to organize themselves in networks to ensure interest representation in a well-balanced democracy and economy or simply feel connected and protected with their friends and family. While such form of network organizations strengthen society, there are also networks created with criminal goals in mind which do the opposite and undermine society. Most empirical studies on networks focus on networks with either of such legal or illegal activities. Baker and Faulkner [34] distinguish legal networks from illegal networks. Participants in illegal and criminal networks conduct their activities in secret: they must conceal the conspiracy from outside 'guardians of trust' such as customers and non-participants inside their own organizations. Various practices and organizational devices are used to protect the secrecy of such a network.

Members may conceal its existence and their involvement in it by limiting face-to-face communication and leaders may be unknown to ordinary members. These goal-directed networks are established with a criminal goal.

Less studies have been carried out on networks involved in legal activities which at a certain point deteriorate. Neither the individual bad apples perspective nor the organizational bad barrels perspective fully explains unethical behavior in organizations. Unethical behavior is foremost a social phenomenon, which needs to be understood by looking at its embeddedness in the structure of social relationships [23]. They argue that interpersonal networks have an important effect on a variety of individual outcomes such as gaining influence and acquiring positions, but more work is needed on network antecedents to understand interaction patterns within and between organizations. This line of thinking calls for a network perspective, both internal and external networks, which are to be understood from their embeddedness in whole networks. Before introducing the far-reaching deterioration of social networks in the form of network corruption, we will first need to understand the two core elements which help to distinguish when a network is a form of social capital or whether it is deteriorating into corruption. These elements are: inclusiveness versus exclusiveness and specific versus generalized reciprocity.

2.5.2   Inclusive Versus Exclusive Outcome

The previous section dealt with the obvious difference between social networks and illegal networks. Some scholars argue that all forms of social capital can be put to unhappy purposes such as suppression or corruption [35]. The advantages that social networks bring about are only one side of the coin. The internal cohesion in networks is achieved at the expense of outsiders (Fukuyama, 2001: 8). In-group solidarity reduces the ability of group members to work together or cooperate with outsiders, thereby often imposing negative externalities on the latter. Fukuyama [30] refers to the corruption this might cause, in particular when the person involved is a public official. In such a case, in-group bonding can cause the cultivation of nepotism (in the interests of family or a group), thus depriving members outside the group from equal opportunity in accessing goods and services. The OECD also points at the risk of relationships which are too inward-looking and fail to take account of what's going on in the wider world [31].

A similar downside to social capital can be seen when differences in education, class and race, result in the underprivileged and disadvantaged losing their connections while the privileged continue to keep their social capital strong [5, 21, 36]. This form of association is considered to operate on the basis of implicit norms, whereby the cultural capital associated with higher education is valued and used to favor some and exclude others. Wuthnow states that social capital may function in an exclusionary way when " it consists of limited networks that provide valuable information to some people but not to others (old boys' networks, for example) or when associations set up expectations about membership that cannot be easily met by everyone" [36]. These networks can be considered homogeneous in the sense that they consist of one or a few centers of influence with a closed group of individuals who are alike and share a certain background (study association, political party, historical roots). Also, they have successfully bridged with others because they work at or represent a wide spectrum of influential institutions or organizations in a given society. Such social networks exist across democratic and societal institutions and are present in the heart of the decision-making processes. This implies that social networks have the potential to enrich, help, support their members at the cost of the outsider. This is reason enough to consider how networks such as those of the local political visualized in images 1 and 2 influence one another and what the impact is in network members and outsiders. Before getting to that question, there is another aspect to consider: the element of reciprocity within networks.

2.5.3 Generalized Versus Specific Reciprocity

The reward of being part of a network is that the network or its members return something of value. This can best be described as the dominant norm being that of reciprocity whereby two forms of reciprocity need to be distinguished. Specific reciprocity is when an individual does something and a particular person does something in return [16]. This exclusive or specific reciprocity (the demand that any favor be reciprocated) tends towards exclusive ways of associating [5]. In this way, particularized reciprocity is seen as being bad for democracy, because it builds on and reinforces group separations. Fukuyama also points to the 'negative externalities' if groups achieve internal cohesion at the expense of outsiders: the larger society in which they are embedded [30]. This favor and return favor can either be a characteristic of a friendly gesture (gift) but equally can be an element of bribe paying.

Generalized reciprocity is noticeable when an individual does something without expecting anything specific back from that other person in the confident expectation that someone else in the network will do something for him at some point [16]. Those who generalize reciprocity will help others not because they expect a particular return, but because they believe that, should they need help in the future, someone else will demonstrate the same spontaneous generosity [5]. In our modern society, this form of generalized reciprocity is of particular value because not every single exchange needs to be balanced. This external focus is complemented by investments in the internal relations, thereby strengthening the social network's collective identity and its effective governance [29]. This norm is derived from the network's purpose and is an intangible element which steers behavior of the individual in the direction of the network's purpose. The question is whether the generalized form of reciprocity linked to networks is a guarantee that opportunities will be available to all. Is it possible that our modern concepts of corruption oversee the potential risks of generalized reciprocity simply because from a legalistic point of view the specific form of reciprocity meets the requirement of the concrete and tangible 'quid pro quo'?

2.5.4   Reinforcing Mechanisms

The general characteristics of social networks and their interplay with the elements of reciprocity and exclusion help in understanding how certain reinforcing mechanisms within social networks increase network deterioration. When individuals organize themselves in networks with people which they identify with, share ideas and beliefs with and these networks over time become the form of social network in which generalized reciprocity becomes the norm and the network becomes inward-looking, this means the network is deteriorating.

The social networks adapt to changing circumstances by attracting, selecting, socializing and retaining individuals to ensure continuity. Because they threaten the group's subculture, particular attention is paid to individuals who cannot be 'turned' [4]. Newcomers may be 'pressured' to encourage compliance with the group's norm. If unsuccessful, the punishment usually escalates and becomes exclusionary; a new-comer is rejected and induced to quit. Exclusionary punishment not only pushes the newcomer from the group, it reaffirms the norms and beliefs of the group by clearly drawing the line between 'acceptable' (corrupt) and 'unacceptable' (non-corrupt) behavior [4]. In this way, corruption is internalized by its members as per-missible and even desirable behavior, and passed on to successive generations of members [4].

Social comparison theory explains how norm formation takes place. Often, indi-viduals go back a long way and the norm formation in their closed networks was influenced by the social comparison among one another. If the networks emerge in the form of closed communities, this implies that the influence by outside attitudes is lacking, resulting in reciprocity being interpreted in a narrow sense; only applicable to the network members. Although Warren [5] looked at norm development within organized collectives such as institutions, he also explains how focusing on individual behavior might detract attention from norm evolvement with corrupt outcomes [5]. Warren [5] and Ashforth and Anand [4] emphasize the importance of acknowledging how norms develop within collectives. In developed democracies, more and more influence and public purposes are attributed to nongovernmental organizations and profit-seeking businesses. The normalization of corruption occurs and becomes an ongoing collective undertaking due to the process of institutionalization, rationaliza-tion and socialization. Corruption becomes a property of the collective and thereby an integral part of the daily conduct to such an extent that individual members may be unable to see the inappropriateness of their behavior [4]. Corrupt individuals often acknowledge their conduct but deny the criminal intent [4]. Most individuals still uphold values such as fairness, honesty and integrity, even as they engage in corrup-tion, that is, until the momentum of a corrupted system provides its own seeming legitimacy. This distances individuals and groups from the moral stance implied by their actions and perhaps even forges a moral turn around, in which the bad becomes good. Networks can create their own norms at odds with the outside world to the point where they become a 'law unto themselves' [37]. Because of this, networks can be quite secret and difficult to examine or regulate in their operation and endeavors.

The reinforcing mechanisms in networks result in the misuse of professional roles for network interests leading to preferential treatment of network members, thereby no longer serving the interests one has to observe when carrying out his role or profession. As such the actual informal role in the network influences the professional role one formally accepted. The individual examples of preferential treatment in such networks do not meet the criteria of the legal definition of corruption because there is no abuse of power in return for a private gain. However, the network dynamics allow improper interests to influence decision-making at the expense of the general interest.

2.6 Network Corruption

2.6.1 Defining Network Corruption

By introducing the idea of 'network corruption', the structure that nurtures the development of corrupt practices receives more attention, and this is a step in the direction of considering corruption as a complex system, the prevention of which requires alternative approaches.

Network corruption concerns individuals having organized themselves in networks that function as a social system in which the interaction of various individuals results in corruption but in which the behavior of the individual is not necessarily corrupt and can best be seen as a form of abuse of power, favoritism, conflict of interest, etc. The real corruption is in the nature of these networks in which network members favor other network members, thereby developing the norm of generalized reciprocity which results in the exclusion of non-network members. This is referred to as "network corruption" (corruption by the network) in addition to the corruption networks (corruption in a network). The recent study on the links between social networks and corruption defines network corruption as: "Informal collective cooperation in which professional roles are misused for network interests to such an extent that the dominant norm is that of generalized reciprocity, leading to the exclusion of others, while the members' awareness of their network is reflected in their common attitude" [38].

Such networks often start off as normal healthy networks and can be seen as a valuable form of social capital in our society. But those features that allow networks to achieve positive results can also have a downside that leads to the deterioration of the networks. The result of the decision-making that has been influenced by undue interests is the core aspect of the corruption, and not the actual exchange of interests. This way, corruption is not something which can only be attributed to an individual. Here corruption is the outcome or result of a (social) process. Instead of assessing the nature of individual actions, this definition seeks to look at the outcome of a process before labeling something as "corruption" [38, 39].

Table 2.1 The features of networks and the indicators when they form social capital or network corruption

Variable	Social capital	Network corruption
An informal collective cooperation	Cooperation	Misuse of professional roles for network interests
Shared interest	Specific reciprocity	Generalized reciprocity
Common attitude	Consciousness of the individuals	Consciousness of the collective
Closed character	Strong bond	Exclusion

The four features outlined in the table help us to see the mechanisms that cause networks to deteriorate. Although the exact tipping point for network deterioration is hard to discover, these features interplay in such a way that certain observable mechanisms start to work. The misuse of professional roles for network interests leads to the preferential treatment of network members, thereby preventing proper professional interests being served. In the process, the actual role in the network takes over from the professional role that had been formally accepted. As a result of the social process of these networks, preferential treatment becomes the norm (a form of generalized reciprocity). This implicit norm steers behavior to reinforce the norm, and network members are become blinded, no longer perceiving that they have a choice between a commitment to the network or to third parties [26]. The norm steers their behavior to such an extent that they act, refrain from acting, decide, and think in the interest of the network collectively, and in this way the network becomes closed, to the detriment of the rights of outsiders. This common attitude is strengthened by the network members increasingly identifying themselves with the network, and its closed nature leads to the absence of internal criticism and correction mechanisms. That, in turn, reinforces all the other features. The mutual reinforcement of these features corrupts the network [38] (Table 2.1).

When talking about network corruption, two aspects need to be distinguished. First of all, network corruption is about individuals being so entangled in a network that their individual conduct can influence one another in a way which is unexpected and results in effects which are unpredictable. This is complex collective acting (corruption by the network). Secondly, network corruption is the enabler of other forms of criminal conduct. Phone hacking, violating public procurement rules, vote buying, leaking of confidential information, fraud and intimidation are interwoven with the corruptive acting (corruption in the network). These crimes are interrelated and they all concern abuse of power, reciprocal mechanisms and acts related to the covering up of these acts.

Network corruption is neither about individual corrupt behavior nor about an entire state of being which is corrupt. Network corruption is the phenomenon which captures the collective conduct in a social network which at a certain point changes from social capital into corruption (emergence). Although decision-making is often influenced in unlawful and improper ways, corruption is in fact a violation of the essence of fair decision-making, whereby decision-making is no longer protected by values of legitimacy, transparency and integrity. The improper interests influence decision-making in a more subtle way. Furthermore, by introducing 'network corruption', the structure in which corruption develops receives more attention, which means a step in the direction of considering corruption to be a complex collective behavior, the prevention of which requires alternative approaches.

2.6.2   Implications of Network Corruption

The main purpose of this chapter was to increase our understanding of the link between networks and corruption. The structure of a network and norm development can be fertile ground for committing collective corruption (corruption in the network). Although this is a risk to consider by social network members and policy makers, the legal provisions and policies available allow the effective prosecution and conviction of this form of corruption. This intentionally committing of corruption together with network members is of a different order than networks in which the norm of reciprocity slowly develops into corruption. Here they start to develop the characteristics of autonomous processes and organizations. This complex phenomenon can best be described as corruption by the network which exists alongside the corruption in the network. Network corruption exists above and beyond the individuals involved. As such, this chapter has followed up on the call by of Ashforth and Anand [4], Zimelis [6] and Vergara ([7], to add new integrated perspectives which help to understand and address systemic corruption in western societies. If networks form the context in which corrupt practices such as bribery become the norm, they are referred to as 'corruption networks'. Although this phenomenon also concerns the link between network and corruption, this article dealt with a new phenomenon: network corruption.

When considering the implications of network corruption, it is foremost essential that we acknowledge that good people can collectively create corruption. Assessing ourselves and each other as 'good apple' or 'bad apple' is no longer valid when trying to prevent network corruption. This requires a different awareness: a network awareness. A network awareness implies that we all are aware that when we participate in networks, we automatically are influenced by these networks. Networks as such form valuable social capital but also have a downside or risk which too often we are not aware of. When wanting to prevent corruption we should have a network awareness in which we also keep an eye on the possible negative consequences of our networks and networking activities. It means we have to make concrete what roles we take on in these dynamic social networks and how we prevent these networks from becoming exclusive and built on generalized reciprocity. In doing so, we must acknowledge that these networks have a strong influence on how we observe ourselves and the network. The indicators in the table distinguish networks which form social capital from those who deteriorate into network corruption. These features can be operationalized in a network tool to assess the character of a person's individual networks and meta networks within a certain context.

Many countries have legal anti-corruption and integrity frameworks. However, most frameworks suffer from loopholes when it comes to assessing the informal processes of networks; they fall short in recognizing network corruption. It is necessary to explain how independent legal instruments, such as rules on public procurement, revolving-door, lobbying and the various integrity policies such as code of conducts and obligation to register gifts and additional functions, not only seek to prevent the individual act of corruption but also corruption as an outcome of collective acting. Often bid rigging, revolving-door constructions and unfair forms of lobbying do not concern an individual seeking influence but are manifestations of a network.

When thinking about the collective behavior of networks resulting in damage, this also requires (re)considering the introduction of collective responsibility. Collective responsibility in addition to individual responsibility has been considered and used in exceptional cases of violence and crimes against humanity. However, in addition to network corruption, there are various other examples of serious societal or global damage caused by informal collectives, such as damage caused to the environment (global warming, loss of biodiversity) or harm to societal health (obesity). Although responsibility in a moral and legal sense starts with the search for who is to blame, this search will soon result in singling out individuals and their contribution to the collective harm. For instance, car owners are to blame for their cars' emissions and the effect they have on global warming, while national governments are blamed for not granting more subsidies for green and sustainable initiatives. These examples of complex collective acting and responsibility have a lot in common with the emergence of network corruption. It is the total sum of the parts which creates the negative impact while its individual parts are not necessarily a wrongful act. As a society we struggle to address responsibility for collective acting. However, if there is a consciousness of the collective this could be a reason to treat networks as a moral actor with similar responsibilities to prevent wrongdoing as legal entities have. If we gain a more overall network awareness and also challenge one another to reflect upon our role within networks and discuss the network' outcome with network members and outsiders, this could be step in the direction of collective responsibility for network outcomes.

The concept of network corruption we offer here intends to be a step towards conceptualizing corruption as a social process helping to lift the rational actor approach and positivist veil that currently hinders the effectiveness of anti-corruption policies in western democracies.

References

1. Slingerland W (2018) Network corruption: when social capital becomes corrupted: its meaning and significance in corruption and network theory and the consequences for (EU) policy and law. PhD thesis, Vrije Universiteit Amsterdam
2. [United Nations Development Programme]. Anti-corruption portal. https://anti-corruption.org/. Accessed 13/01/21
3. [Transparency International]. What is corruption? https://www.transparency.org/en/what-is-corruption. Accessed 20/01/21

4. Ashforth BE, Anand V (2003) The normalization of corruption in organizations. Res Organizat Behav 25:1–52
5. Warren ME (2004) What does corruption mean in a democracy? Am J Polit Sci 48(2):328–343
6. Zimelis A (2020) Corruption research: a need for an integrated approach. Int Area Stud Rev 23(3):288–306
7. Vergara C (2021) Corruption as systemic political decay. Philos Soc Criticism 47(3):322–346
8. Klitgaard R (1998) Controlling corruption. University of California Press, Berkeley
9. Warburton J (2013) Corruption and anti-corruption. In: chapter Corruption as a social process: from dyads to networks. ANU Press, Canberra, pp 221–237
10. Pinto J, Leana CR, Pil FK (2008) Corrupt organizations or organizations of corrupt individuals? two types of organization-level corruption. Acad Manag Rev 33(3):685–709
11. Johnston M (1996) The search for definitions: the vitality of politics and the issue of corruption. Int Soc Sci J 48(149):321–335
12. Nielsen RP (2003) Corruption networks and implications for ethical corruption reform. J Bus Ethics 42(2):125–149
13. Huberts LWJC, Six FE (2012) Local integrity systems. Publ Integr 14(2):151–172
14. Brass DJ, Butterfield KD, Skaggs BC (1998) Relationships and unethical behavior: a social network perspective. Acad Manag Rev 23(1):14–31
15. Owens LA, Young P (2008) You're hired! the power of networking. J Vocat Rehabil 29:23–28
16. Putnam RD (2000) Bowling alone: the collapse and revival of American community. Simon & Schuster, New York
17. Provan KG, Fish A, Sydow J (2007) Interorganizational networks at the network level: a review of the empirical literature on whole networks. J Manag 33(3):479–516
18. Douglas S, Sónia A, Balázs É, Tomasz K (2011) Public policies and investment in network infrastructure. OECD J: Econ Stud 1:2011
19. Provan KG, Lemaire RH (2012) Core concepts and key ideas for understanding public sector organizational networks: using research to inform scholarship and practice. Public Adm Rev 72(5):638–648
20. Vilkas M (2014) The role of emergent networks in a planned change of organizational routines. Transform Busin Econ 13(2):188–206
21. Granovetter MS (1973) The strength of weak ties. Am J Sociol 78(6):1360–1380
22. Meadows DH (2011) Thinking in systems. A primer. Earthscan Ltd, London
23. Brass DJ, Galaskiewicz J, Greve HR, Tsai W (2004) Taking stock of networks and organizations: a multilevel perspective. Acad Manag J 47(6):795–817
24. Turner JC, Hogg MA, Oakes PJ, Reicher SD, Wetherell MS (1987) Rediscovering the social group: a self-categorization theory. Blackwell, Oxford
25. Davis JB (2014) Social capital and social identity. Routledge, London, p 290
26. Heffernan M (2011) Willful blindness: why we ignore the obvious at our peril. Walker Publishing Company, New York
27. Baumeister RF, Leary MR (1995) The need to belong: desire for interpersonal attachments as a fundamental human motivation. Psychol Bull 117(3):497–529
28. Gabriel S (2021) Reflections on the 25th anniversary of Baumeister and Leary's seminal paper on the need to belong. Self Ident 20(1):1–5
29. Adler PS, Kwon S-W (2000) Social capital: the good, the bad and the ugly. Butterworth-Heinemann, pp 89–115
30. Fukuyama F (2001) Social capital, civil society and development. Third World Quart 22(1):7–20
31. OECD (2001) The well-being of nations: the role of human and social capital. OECD Publishing, Paris
32. Dasgupta P, Serageldin I (2000) Social capital: a multifaceted perspective. The World Bank, Washington
33. Burt RS (2000) The network structure of social capital. Res Organ Behav 22:345–423
34. Baker WE, Faulkner RR (1993) The social organization of conspiracy: illegal networks in the heavy electrical equipment industry. Am Sociol Rev 58(6):837–860

35. Szreter S, Woolcock M (2004) Health by association? social capital, social theory, and the political economy of public health. Int J Epidemiol 33(4):650–667
36. Wuthnow R (2002) Loose connections: joining together in America's fragmented communities. Wiley, Hoboken
37. Granovetter MS (1992) Economic action and social structure: the problem of embeddedness. Westview Press, Oxford, pp 22–45
38. Slingerland W (2019) Network corruption: when social capital becomes corrupted: its meaning and significance in corruption and network theory and the consequences for (EU) policy and law. Eleven Publishing, Amsterdam
39. Slingerland W, de Graaf G (2020) The Netherlands: an impression of corruption in a less corrupt country. Edwar Elgar, pp 376–387

Chapter 3
Network Controllability Metrics for Corruption Research

Philip C. Solimine

Abstract This chapter will discuss political corruption in a legislative social network using the tools of network control theory. We aim to cultivate an understanding of the mechanisms by which corrupting actors can perturb nodes' behavior at certain points within a larger social system, and harness the natural magnification of these perturbations to drive the network to a more desirable state. In other words, we investigate how social capital is harnessed to amplify the effects of corruption using a social influence network or hierarchy. We provide a brief overview of control-theoretic metrics that may prove useful to researchers in identifying high-risk areas of legislative or other political social networks—areas which are particularly vulnerable to the spread of misinformation, and thus particularly valuable to the corruption of social dynamics.

3.1 Introduction

The dynamics of political opinions, particularly within a deliberative assembly, play a crucial role in the evolution of a society's legislation. The opinions of a politician can change and adapt over time, and co-evolve with others through discourse; the opinions they form corresponding directly with the types of laws and regulations that these politicians will sponsor or vote in favor of.

An unfortunate byproduct of the inherently connected nature of social interaction is that the actions of a small number of corrupt individuals can drastically affect the trajectory of the entire social dynamics. This chapter will address corruption as a control problem in a social network, in which a society of individuals discuss and exchange opinions dynamically over a complex network structure. We hope that this will provide a basis for researchers interested in the study of "quid-pro-quo" corruption as an intervention or control within the dynamics of a social network. We are motivated by the sociological observations of [15], in which Mark Granovetter argues

P. C. Solimine (✉)
Florida State University, Tallahassee, FL, USA
e-mail: psolimine@fsu.edu

that "The ability to effectively corrupt the administration of some substantial activity requires corruption entrepreneurs to be masters of social network manipulation."

In this chapter, we will focus only on a very specific type of corruption, which is "quid-pro-quo", closely related to studies of the spread of misinformation in social networks. In this type of corruption, an outside agent or agency ("outside" meaning that they are not a part of the social network in question) called the *corrupter* or *attacker* makes a monetary transfer or payment to a node in the network (a politician) to misrepresent or *perturb* their opinion on some topic. Since the individuals in the network are exchanging opinions with each other, these perturbations can have effects which propagate through other nodes in the network, moving all of their opinions toward states which are more desirable to the attacker.

We begin with an introduction of the different models used to describe the dynamics of opinion exchange and learning in social networks. Then we introduce corruption as a control-theoretic optimization problem. Finally, we discuss what happens when nodes are given the freedom to change their social ties over time, and the emergent properties of endogenous networks, in which node's incentives to build an effective network are afflicted by corruption.

3.2 Motivations

If nodes primarily choose their incoming influence links, then it is easy for them to manipulate a specific measure called control capacity [31], which is the fraction of minimum driver sets in which they are contained. A large number of nodes with high control capacity means a network that is difficult to control. If nodes form outgoing links, or are primarily concerned with increasing their control centrality—the number of other individuals that they can directly or indirectly control—then they should attempt to become connected in the strongly connected component that is at the top of the hierarchy, preferentially forming outgoing influence links to other individuals with long matching paths in order to command power over a larger fraction of the network. If this were the case, we should expect to see the network become *easier* to control as corruption increases due to increasing lengths of matching paths [38]. Intuitively, an individual with high control *centrality* commands high power in the network in that they can control a large fraction of others, while one with high control *capacity*, while may not control many others, is more important because they are harder to influence from elsewhere in the network.

In prior work [36], we have found that the fraction of driver nodes required to control a political social network increases with measures of corruption perception across a panel of European parliaments. This provides evidence in favor of control capacity as a measure of dominant importance to identifying high-risk areas of the network for corruption and the spread of misinformation in social networks. In other words, individuals in the network who wish to accept bribes in exchange for spreading misinformation or misrepresenting their own position on an issue appear to focus on ensuring they are not indirectly controlled through others, rather than focusing on controlling many others themselves.

These results suggest that as corruption increases, the network can become harder to control. This effect was named hierarchical congestion [36]. This is particularly relevant in light of other work on the controllability of social networks and its relation to corruption. In [7] the authors examine the controllability of the Brazilian federal police network, and find that a minimum driver set may consist of only roughly 20% of the network nodes. They highlight the importance of easy control in a properly functioning network, and claim that easy controllability is valuable because it can allow for beneficial norms, such as a resistance and moral objection to corruption, to spread easily through the network. This reasoning extends to networks in a deliberative assembly, and in particular, one in which the nodes are representatives of large communities. If a number of communities strongly wish to enact change in the political system, a well-functioning democracy should enable them to enact such change effectively. Thus, when such control is coupled with monetary incentives in the case of "quid-pro-quo", the nodes in the network appear to change their linking strategies in a way that is detrimental to the effectiveness of the democratic process.

With this in mind, it is important to ensure that our discussion is properly motivated. Controllability of a social network system is not inherently a bad thing—it can allow for an effective democratic system and an ability to encourage new norms. Thus, while control theoretic centrality metrics point to a set of opinion leaders that are at high risk for corruption, these same features can also make them invaluable to anticorruption efforts [7]. This adds another layer of importance to the finding that corruption is associated with poor controllability [36], and will hopefully also convince the reader that these control metrics are far from a guaranteed mechanism by which to identify corrupt agents. Rather, it is only when this type of influence is overly incentivized through corruption opportunities that it becomes problematic.

Notable recent work [11] has also discussed adversarial perturbations to a more advanced social learning model called the Friedkin-Johnsen model [10]. The Friedkin-Johnsen model addresses what is often seen as a flaw of the DeGroot model, which is that its dynamics inevitably converge to a consensus state in the limit, in which all opinions are equal. It builds on the basic DeGroot model to allow for persistent disagreements, by considering each node as being anchored to an immutable initial opinion as well as considering the opinions of their neighbors.

Adversarial perturbations are a concept developed in machine learning and deep, feedforward neural networks. They are based on a finding that deep network-based classifiers can often be easily fooled into misclassifying data by tiny, strategically placed perturbations to their inputs [33]. In [11], the authors extend this theory to opinion dynamics, discussing perturbations that induce discord in the converged final state of opinions. They show that this optimal perturbation can be designed based on the spectrum of the graph Laplacian. The primary difference between this type of adversarial perturbation and control theory is twofold; adversarial perturbations may be injected to the states of *any* node in the network, and *only in the first time period*. This would be roughly analogous to the case in which each node is a driver, capable of receiving an independent input signal, but the input signals are restricted to be zero for all time periods after the one *leading* to the initial state x_0. Further, they are primarily interested in sowing the seeds of disagreement in the limiting distribution of states, rather than reaching a specific goal state by a specific finite time threshold. This is consistent with other prior work on control in social network opinion dynamics [32], but may not be ideal for the study of corruption in which reaching a goal by a time threshold (such as before a vote) plays a more important role to the corrupter than does the limiting distribution of beliefs or opinions.

3.3 Graph Preliminaries

A social influence network can be viewed as a graph $G = (N, \mathcal{E})$, consisting of a set of n vertices (nodes) N, connected by a set of edges or links $\mathcal{E}$. Each node i in N also possesses a state variable $x_i \in \mathbb{R}$ which represents their opinion on some topic of discussion. These are collected in a state vector $x \in \mathbb{R}^n$. In a social network, we take the nodes in N to be the individuals participating in the society, and the edges (or *links*) in $\mathcal{E} \subset N^2 \times \mathbb{R}$ to be an ordered pair of nodes which represent friendship or interactions, coupled with a real-valued weight for the interaction w_{ij}. In other words, each individual forms their opinion as a weighted average of their neighbors'. The graph structure can be summarized by a weighted, row-stochastic adjacency matrix A, an $n \times n$ matrix, with entry $A_{ij} = 0$ if there is no link from j to i, and $A_{ij} = \frac{w_{ij}}{\sum_{j=1}^{n} w_{ij}}$ otherwise. In the case that $\sum_{j=1, j \neq i}^{n} w_{ij} = 0$, (if a node is a root with no incoming links), then they are assigned $w_{ii} = 1$, a self-loop, to ensure that their state or opinion will remain constant unless they are actuated. Such nodes may be referred to as stubborn. Introducing this stubbornness is a convenient method to deal with the directedness of the network; it does not affect the control capacity or control centrality, but does affect some efficiency metrics by making the node more expensive to control.

3.3.1 Social Learning and Opinion Dynamics

Complex networks can be a powerful tool for modelling the diffusion of information and opinions through a social structure. Studies of opinion dynamics in networks are often centered around the theme of political discussion [18, 19], and for obvious reasons; the features of the democratic process mean that political discussion in social environments plays a substantial role in the evolution of laws and regulations which directly affect the progression of a society and its norms. Only recently has the literature begun to examine the role of network formation and endogeneity in such environments, and such studies are sparse [3, 25]. This is not surprising, as network formation is computationally difficult to analyze unless subject to strict assumptions on the process [6].

Much of the study of learning and opinion dynamics in social networks is concerned with the consensus—an equilibrium state in which all individuals agree on the same opinion—and the accuracy of equilibrium beliefs [4, 13]. The most common model, due to its relative simplicity, is called the DeGroot model. In a DeGroot model—named for the work of DeGroot [8]—each node simply considers the opinions of those with which they are linked, and updates their own opinion as a convex combination of their neighbors' opinions.

$$x_i(t+1) = \sum_{j=1}^{n} \frac{w_{ij}}{\sum_{j=1}^{n} w_{ij}} x_j(t) \tag{3.1}$$

These equations can be written in matrix-vector form as:

$$x(t+1) = Ax(t) \tag{3.2}$$

Given the initial state of opinions $x_0 \in \mathbb{R}^n$, the opinion at discrete time period $t \in \{0, 1, \ldots, T\}$ is given by:

$$x(t) = A^t x_0 \tag{3.3}$$

A crucial feature of this process, is that it can represent the naïve best-response dynamics (or simultaneous gradient descent) in a *pure coordination game* [4, 14]. In this context, a pure coordination game is a game in which the nodes optimize an objective that is negative quadratic in the difference between their own state and that of their neighbors. Individuals have the following utility:

$$U_i(A, x) = - \sum_{j=1}^{n} w_{ij} (x_i - x_j)^2 \tag{3.4}$$

We assume that player i is following the best-response dynamics in discrete time intervals. These best-responses arise from the following first order condition:

$$\frac{\partial U_i}{\partial x_i} = 2 \sum_{j=1}^{n} w_{ij}(x_i - x_j) \equiv 0 \tag{3.5}$$

The linear DeGroot dynamics (3.1) are easily found by algebraic manipulation of this first-order condition giving best responses. Therefore, if a player in the network is myopically updating their opinions to satisfy this first-order condition (themselves intrinsically assuming that all other players' opinions will remain the same[1]), then the DeGroot learning process can be viewed as best-response dynamics of this game.

---

[1] The fact that in the best-response dynamics, players do not take into account how others in the network will update their opinions in each round is what makes these dynamics *myopic* or *naïve*.

3.3.2 Control Theory

Concurrently, a largely disjoint but closely related literature has been developing across engineering disciplines called control theory. Control theory can be loosely described as the study of how dynamical systems can be supplemented with controllers, in a way that drives the system to a more desirable state.

Conveniently, the majority of DeGroot-type models of social learning fall into the category of linear systems, for which closed-form results can be obtained to describe optimal control. A large body of literature addresses the controllability of linear systems, and in particular how it can be accomplished at minimum cost [26, 28, 35]. In particular, in [35] a number of widely-used metrics are described which aim to quantify the energy or difficulty required to drive the dynamics of a linear system.

Formally, control in a discrete-time linear (time-invariant) system with m control inputs is described by the following system of difference equations:

$$x(t + 1) = Ax(t) + Bu(t) \tag{3.6}$$

In this system, the $m \times 1$ vector u_t lists control input signals, and the $n \times m$ matrix B maps these input signals to the states of specific sets of nodes. In other words, the matrix B allows for perturbations to be injected into the states of a certain set of nodes in any time period.

There is an interesting nuance in the way that the matrix B is described. That is, we can restrict the columns of B to those of the canonical basis. In other words, we can place a restriction on the columns of B so that exactly one element of each column is 1, and the rest of elements in the column are 0. This special case corresponds to the situation in which each input signal is sent to exactly one node of the network. In this case, we would refer to the actuated nodes $i \in \mathcal{N}$ for which there exists a j with $1 \le j \le m$ such that $B_{ij} = 1$ as *leaders* or *control nodes*. If the network is controllable from a given set of leaders, and there is no smaller set of control nodes that could fully control the network, then the set is called a minimum driver set [30], and the nodes are referred to as *driver nodes*.

Alternatively, if the restriction that exactly one element of each column of B is nonzero is relaxed, so that an input signal can be sent to multiple nodes simultaneously, then the problem can be viewed as augmenting the network with a set of new vertices. These actuator nodes resemble the original nodes in the network. They connect to a series of other nodes, but do not accept any incoming influence links from other nodes or actuators in the system; rather, their states are determined entirely by their corresponding input signal. Actuator placement of this type can thus be viewed as the strategic positioning of a number of forceful agents into the network, a problem whose effects are studied in detail by [2] with a focus on the spread of misinformation.

We call a network *controllable* if, for *any* desired final state $x_f \in \mathbb{R}^n$, inputs $u(t)$ can be designed which drive the system to x_f. In the case that the network weights are known, or estimated with a high degree of accuracy and precision, *exact controllability* [41] can be used to define whether a network is controllable. Exact controllability is based on a condition for controllability called the Popov-Belevitch-Hautus (PBH) condition [17], which is equivalent to the Kalman condition. The PBH condition holds that the system is controllable if and only if for all complex-valued constants $c \in \mathbb{C}$:

$$\text{rank}([cI - A, \ B]) = n \tag{3.7}$$

The alternative approach, known as structural controllability, relies on a matrix called the controllability matrix of the system. We can define a matrix C_T (the *T-steps controllability matrix*) as:

$$C_T = [B, \ AB, \ A^2 B, \ \ldots, \ A^{T-1} B] \tag{3.8}$$

where we have used the comma to denote horizontal concatenation. Intuitively, a vector x_f falls into the span of the T-steps controllability matrix only if a set of signals $u(t)$, input to the system between time periods 0 and $T - 1$ can be used to drive the dynamics to reach that vector. For example, if $T = 1$, then $C_1 = B$, so the state must fall in the range of the columns of B. If $T = 2$, however, then $C_2 = [B, \ AB]$; the controller now has two time periods to inject perturbations. Input signals injected to the state in the first period are amplified by the natural network dynamics, ideally leading the network to a state from which x_f can be reached through a perturbation proportional to the columns of B.

This leads to a famous result called the Kalman controllability rank condition [23]. That is, the system (3.6) is *controllable*, so that any vector x_f in the state space X can be reached in exactly T steps, if and only if the T-steps controllability matrix C_T is of full (row) rank. As T grows, control can rely more and more on the natural autonomous dynamics of the system, until $T = n$. For all $T \geq n$, the space of reachable states is exactly the range of $C \equiv C_n$. For this reason C is simply referred to as the *controllability matrix* of the system.

3.3.3 Optimal Control

Along with exactly which states are theoretically reachable from a given starting state and control input configuration, it is important to consider the costs that are required to drive the system. These costs are typically framed in terms of the input energy [26, 28, 39]. The energy cost is given by the sum of squared norms of the vector u of input signals. Intuitively, this convex cost makes large input signals more costly than small ones, with the cost of additional energy increasing more for a signal that already has a large energy cost. Formally, the input energy cost is given by:

$$E_T(u) = \sum_{t=0}^{T-1} \|u(t)\|_2^2 = \sum_{t=0}^{T-1} u(t)^\top u(t) \tag{3.9}$$

It is reasonable to assert that the cost is convex in the corruption case, since large instantaneous shifts in opinion may be suspect and thus come at a higher risk for the parties involved. This energy cost is convenient, however, not only because it is realistic in imposing a cost that is convex in the magnitude of each perturbation, but also because it can be used to derive a closed-form solution for the optimal control signal which drives the state to x_f in T steps at the minimum cost. Given the control input schematic B, the unique, optimal, open-loop control input for any time period $0 \le t \le T - 1$, (of course, since T is the terminal period, no more control input can be injected at this point), is given by:

$$u_t^*(x_f) = B^\top (A^\top)^{T-t-1} W_T^{-1}(x_f - x(T)) \tag{3.10}$$

Here, we have used W_T to denote a matrix, called the *Controllability Gramian* of the system, which takes the following form [35]:

$$W_T = C_T C_T^\top = \sum_{t=0}^{T-1} A^t B B^\top (A^\top)^t \tag{3.11}$$

The Gramian is guaranteed to be positive definite (and thus invertible) whenever the system is T-steps controllable under the input schematic B. Further, it allows a closed form solution for the minimum energy required to control the network system B from state x_0 to state x_f in exactly T steps. This cost is given by:

$$E_T^*(x_f) = (x_f - A^T x_0)^\top W_T^{-1}(x_f - A^T x_0) \tag{3.12}$$

Due to its appearance in this expression for minimum energy, the Gramian matrix, or variations of it, are popular for constructing descriptive metrics for the efficiency of a driver set.

3.4 Controllability Metrics

There has been much work [29, 34, 35, 37] aimed at developing metrics that can inform effective input placement and evaluate the importance of various driver nodes to efficient network control. These efforts can largely be broken up into three categories. The first is at the network level. Metrics that describe, for example, the minimum number of driver nodes or actuators that are required to control a network, fall into this category.

Secondly, and perhaps most relevant to the study of corruption in a social networks, there is a growing interest in metrics that describe the importance of an *individual node* to a driver set or input schematic. The aim of these metrics is to build input schematics that are effective and efficient. Thus, they should be of particular importance to the study of corruption—in helping to understand exactly *where* in the network this type of corruption and misinformation spread is most valuable to the corrupter, and thus which areas of the network are at the highest risk for interference.

The final category is metrics that evaluate the effectiveness of a driver set or control input matrix B as a whole and within the network environment. That is, given a valid control input placement, features such as bounds on the control energy required to reach an arbitrary state, or the volume of the ellipsoid describing states that are reachable with a single unit of input energy can give an idea of the efficiency of the schematic. Many of these metrics can be extended to the case of the individual node to give an idea of how important an individual node is to attaining efficient controllability.

3.4.1 Network-Level Controllability Metrics

There are two general approaches to measuring the controllability of a network system as a whole. These are called *exact* [30] and *structural* [41] controllability. Structural controllability is arguably more suited to the study of corruption in social networks, because usually in the study of real-world networks we will not know the exact edge weights associated with each link. Rather we will only have, at best, some noisy estimates of the link weights. The controllability of a network, roughly speaking, is measured by the minimum number of control inputs (linearly independent columns of B) that are required in order for a suitably designed control schematic to have the ability to satisfy the Kalman or PBH conditions.

The minimum rank that B can be, in order for the PBH condition to be satisfied, can be derived as the maximum geometric multiplicity of an eigenvalue of the system adjacency matrix A. In other words, the minimum number of control inputs or drivers n_D required to attain exact control inputs of the system is given by:

$$n_D = \max_{\lambda \in \Lambda(A)} \mu(\lambda) \qquad (3.13)$$

where we have used $\Lambda(A)$ to denote the spectrum (set of eigenvalues) of A, and $\mu(\lambda)$ to represent the geometric multiplicity[2] of the eigenvalue λ.

---

[2] The geometric multiplicity of an eigenvalue is the rank of its associated eigenspace, or the number of linearly independent eigenvectors it is associated with. If the network matrix A is diagonalizable (which is the case if the network is undirected), then the geometric multiplicities of eigenvalues are equal to their algebraic multiplicities, or the number of times they appear as a root of the characteristic polynomial.

Intuitively, the first part of the matrix in the PBH condition $(cI - A)$ is guaranteed to be rank-degenerate for all $c \in \Lambda(A)$, since the eigenvalues are precisely taken as the values which are the roots of the determinant of this matrix and thus the points where it is low-rank. In order for the concatenated matrix to be full rank, there must be at least enough linearly independent columns in B to account for this rank degeneracy.

The structural controllability approach was pioneered by [27], and was applied to network systems by [30]. This approach can be used to find minimum driver sets in environments with a known, directed network structure, but in which there is some uncertainty about the weight of each edge and thus about the spectrum of the system adjacency. A review of early work in structural controllability can be found in [9]. In structural controllability, the nonzero elements of the adjacency are taken as free parameters without any specific prescribed value. Importantly, the controllability results obtained from the structural approach can be shown to be hold for most values of the parameters, except in some zero-measure special cases.

The advantage of the structural approach is that it can be accomplished via a simple and tractable algorithm based on maximum matchings of an associated bipartite graph. This is introduced in [30] and explained in far more detail in [31]. The algorithm begins by creating a bipartite representation of the network. In the bipartite representation, one set of nodes is called the incoming set, and the other is the outgoing set. Each set of nodes in the bipartite graph (incoming and outgoing) contains one node representing each node of the original network. In a bipartite graph, links only exist between node sets, but not within them. The original network is then redrawn in its bipartite representation; if a link exists from node i to node j in the actual network, an undirected link is drawn in the bipartite graph between node i in the outgoing set and node j vertex in the incoming set.

After the bipartite representation is constructed, a maximum matching of the new bipartite graph must be found. A matching in a bipartite graph simply describes a set of edges with disjoint vertices. In other words, it is a selection of edges from the bipartite graph, selected so that no edge in the matching shares a node with any other edge in the matching. Further, in order for this matching to be *maximum*, it must be of the largest possible cardinality; containing the largest possible number of nodes from each set, so that there are no possible matchings that contain more nodes.[3]

Once a maximum matching in the bipartite representation is obtained, a minimum driver set for structural controllability is computed as the number of unmatched nodes in the in set. These nodes correspond with nodes in the original network who do not have an incoming link through which they can receive the input signal indirectly. Rather, in order for these unmatched nodes to be controlled, they must be controlled directly by an input signal through a connection in the input schematic.

---

[3] Algorithms for finding maximum matchings in a bipartite graph are well-established. For more information and a comprehensive overview of algorithms for finding maximum bipartite matchings, see [24].

3.4.2 Node-Level Controllability Metrics

An easy way to visualize how the maximum bipartite matching works to find a minimum driver set is by understanding the hierarchical structure that is generated by a control input schematic. This hierarchical structure is described in detail by [31], where authors develop a metric called *control centrality* which helps to shed light on this hierarchical structure.

Control centrality roughly captures the power of an individual node in controlling the entire network. The control centrality of node i is given by the generic rank of an individual controllability matrix $C^{(i)}$. This individual controllability matrix is constructed according to (3.8), but with the control schematic set to include only node i. In other words, B is taken to be $B = B^{(i)} \equiv e_i$, where e_i is the ith column vector from the canonical basis—with a 1 in the ith position and 0's elsewhere. Generic rank refers to the fact that the weights of the network links are unknown, so a maximum rank is computed. Notably, as with the structural controllability, the generic (maximum) rank is equal to the actual rank of $C^{(i)}$ in all but some pathological (measure zero) special cases of the network weight parameter values.

Control centrality describes the maximum matching path that can be drawn through the network from an individual node. If node i is selected as a control input, their control centrality is the generic dimension of the associated controllable subspace of the system. The generic rank of this matrix gives the length of the longest "cactus" shaped matching path that can be drawn beginning at node i. A cactus-style matching path refers to the potential shape of the maximal structure that a node can control. This structure consists of a single directed path (called a "stem") along with any loops that can be directly reached off the stem (called "leaves"). An example of a cactus matching path is shown in Fig. 3.1.

Another individual-level metric is called *control capacity* [21]. The central idea behind control capacity is that the maximum bipartite matching of a digraph is generally not unique. Indeed, there may be several maximum matchings, each lending themselves to a potentially different minimum driver set. Control capacity measures the proportion of minimum driver sets which contain each individual. For example, if there is a node which absolutely must be controlled in any minimum driver set, then their control capacity is equal to one. On the other hand, if a node never appears unmatched in a maximum bipartite matching, then they are never a driver node and their control capacity is zero. If a node appears in some, but not all, minimum driver sets, then they are considered intermittent, and their control capacity is between zero and one. Formally, control capacity can be seen as the probability that a node is selected as a driver node if a uniformly drawn maximum matching is used to select a minimum driver set.

In [21], the authors also develop a method for uniformly drawing minimum driver sets. This is crucial because the number of potential maximum matchings can easily become extremely large and render a direct computation of control capacity infeasible. A method of uniformly drawing from the set of maximum matchings enables Markov-chain Monte Carlo methods for estimating control capacity. The authors

P. C. Solimine

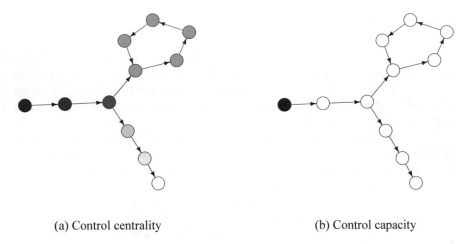

(a) Control centrality   (b) Control capacity

Fig. 3.1 Node-level controllability metrics for a "cactus" style matching path. Nodes with a dark color correspond with those of high centrality, as measured by control centrality (**a**) or control capacity (**b**). In this case, the dark node on the left-hand side is the most central in both cases. In (**a**), the leftmost (darkest colored) node has a control centrality of 11, since it can fully control the entire matching path. Subsequent nodes can control fewer nodes, since they cannot control others that are higher up in the hierarchy but can control those that they connect to either directly or indirectly through a directed path. Finally, the bottom node (white) has the lowest control centrality of 1, because it cannot control any other nodes than itself. In (**b**), nodes are highlighted by control capacity. Since the entire network can be controlled by a single node (by design), the control capacity of the darkest node on the left is 1 (it appears in 100% of minimum driver sets) while the other (white) nodes all have a control capacity of 0 (they never appear in a minimum driver set)

use this to find control capacity measures for a number of real-world and model networks, and find that control capacity depends almost entirely on the distribution of in-degrees (nodes with high control capacity correspond with nodes of low in-degree), and appears to be virtually independent of a node's out-degree.

Importantly, through testing in real-world and model networks, the authors find another surprising but important result; which is that driver nodes tend to "avoid" hubs. This is because a matching path can only take one link associated with each node, so having a large number of links is not necessarily a benefit in terms of control position. In particular, if the hub has even a single incoming influence link, then the node that influences them can indirectly control the rest of their longest matching path or any of the outgoing matching paths that they are connected to.

The simplest example of this phenomenon is in a directed star network as shown in Fig. 3.2. In this graph structure, there is a single hub with outgoing connections to all other nodes (and there are no other links in the network). While the hub should always be selected as a driver node, their longest matching path contains only 1 node and therefore their control centrality is only 1, so while they have a high control capacity, all but 1 of the other nodes must also be selected as drivers so control is highly distributed through the network. Further, while the hub is always a driver in

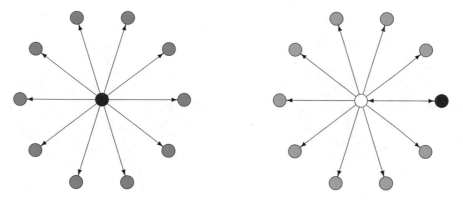

(a) Control capacity in a directed star   (b) Control capacity with an added edge

Fig. 3.2 Distribution of control capacity in a directed star network. This figure highlights how small changes in the topology of a network can lead to large changes in the distribution of control capacity (and thus, substantial changes in the controllability of the network). Note the relative lack of importance of a node's out-degree in determining control capacity. In **a**, the center node has a control capacity of 1, since it has no incoming links but influences all nodes in the network, it must be controlled in any minimum driver set. Its control centrality, however, is only 2. This is because once a node has been matched with another, it cannot be matched again with other nodes. A minimum driver set in **a** thus consists of the single node in the core, along with nine of the remaining ten nodes. Since the center node can be matched with exactly one other node, there are ten minimum driver sets, one for each peripheral node being matched with the core and thus not appearing in the set. In **b**, the addition of a single incoming link to the center node immediately renders it unnecessary as a driver, giving it a control capacity of 0 (although its control centrality remains unchanged). This is because it can now be controlled externally from a peripheral node (the rightmost node), with a matching path then extending to another peripheral node, generating a matching path of length three, and thus requiring one fewer driver. By the same logic as before, each of the other peripheral nodes appear in all but one driver set. Since there are now only nine of them, each peripheral node now has their control capacity slightly reduced from $\frac{9}{10}$ in **a** to $\frac{8}{9}$ in **b**

the directed star, adding even a single incoming link to the central hub causes its control capacity to fall to zero, as the incoming link will allow it to be controlled indirectly.

Situations in which a large proportion of the nodes have nonzero control capacities are referred to as networks that enable distributed control; while networks in which control capacity is zero for most of the nodes are considered networks with centralized control. A further area of interest in the field is the bimodality of control. That is, very small perturbations of the network topology can have enormous effects on the distribution of control measures. This topic is discussed in detail by [22]. In other related work, [38] shows how minimal perturbations can be made to the structure of the network in order to optimize its controllability, which is accomplished by linking the end points of matching paths with the starting points of others.

The cactus matching path in Fig. 3.1 is the most straightforward example of a topology which enables centralized control, since it has a unique minimum driver set

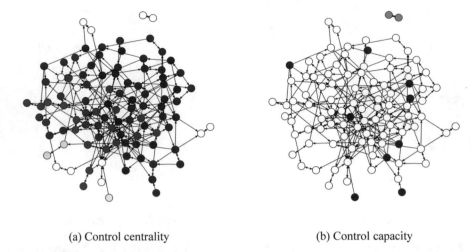

(a) Control centrality   (b) Control capacity

Fig. 3.3 Node-level controllability metrics for the 108th Senate of the United States of America (2003–2005). These network data were obtained from [5], with links drawn by weighted propensity to cosponsor [16], thresholded at 0.15. Notably, this network can be controlled from 10 nodes. Since the two nodes in the upper right are not connected to the rest of the network, but share a bilateral connection, either of them may be selected to drive themselves and the other, leading to two possible driver sets. This is a relatively low number of driver sets, so we may consider this "centralized" control (as opposed to "distributed" control, in which a large number of nodes have nonzero control capacity), using the terminology of [22, 40]. Nodes are colored by their centrality metrics, with darker nodes being more central and lighter ones less central. For control centrality (**a**), the maximum for an individual node is 38 (dark purple) and the lowest is 2 (white) because there are no sinks, so every node has an outgoing connection to at least one other. For control capacity (**b**) the dark nodes are always drivers and thus have a control capacity of 1. White nodes have a control capacity of 0 because they never appear in driver sets, and the two intermittent nodes in the top right have a control capacity of $\frac{1}{2}$ because they form a cycle, so either of them can be selected with equal probability if driver sets are drawn uniformly. Summary statistics for these measures can be found in Table 3.1

consisting of a single root node. The directed star from Fig. 3.2, on the other hand, enables a distributed control configuration because every single node in the network is a part of a minimum driver set. Using cosponsorship network data from [5], we computed the control capacity of senators from the 108th Senate of the United States of America. The results of this computation are displayed in Fig. 3.3, and summary statistics in Table 3.1. Notably, a minimum driver set in this network consists of 10 nodes, and control is centralized. Apart from two nodes which themselves form a component cycle and thus have a control capacity of 0.5, the minimum driver set is unique and thus all driver nodes in the largest strongly connected component have a control capacity of 1.

Table 3.1 Sufficient statistics for node-level controllability and efficiency metrics for the 108th Senate of the United States of America (2003–2005). This network consists of 100 nodes and 260 edges, with centralized control and a minimum driver set size (computed by exact controllability) of $n_D = 10$. It was constructed by weighted propensity to cosponsor [16] thresholded at 0.15, and compiled by [5]. In this case, we have taken cosponsorships to indicate incoming influence from the bill sponsor

Metric	Minimum	Maximum	Average	Median	Std. Dev.
Control capacity	0.00	1.00	0.10	0.00	0.29
Control centrality	2.00	38.00	29.94	35.00	11.80
Node-to-network p_i	1.03	823.14	37.73	1.95	129.57
Network-to-node q_i	14.36	101.00	41.22	37.73	27.58
Non-normality (differenced) $r_{\text{diff},i}$	−98.88	722.14	0.00	−21.52	113.98
Non-normality (ratio) $r_{\text{quot},i}$	0.02	8.15	34.71	0.08	1.27

3.5   Efficiency Metrics

While controllability metrics give a solid understanding of *how many* individuals must be controlled in order to fully capture the state of opinions in the network, they are agnostic with respect to *which* of the potentially many possible driver sets is "best". This is where efficiency metrics shine. Using efficiency metrics to design a control input schematic allows the designer to choose not only a driver set that can be guaranteed to get them where they want to go, but also to ensure that they can do so at the lowest possible cost.

Crucially, even fully controlled network systems may exhibit practically unreachable states when control energy is limited [35]. Practically unreachable states are those which technically fall into the controllable subspace of the system, but for which the minimum energy of a control input signal that would drive the system to this state is beyond any reasonable limit. Efficiency metrics can thus be used to understand and design efficient and effective control input schematics.

3.5.1   Schematic-Level Efficiency Metrics

Many efficiency metrics are based on the system Gramian W_T given by (3.11), which is useful due to its appearance in the minimum energy function (3.12). These metrics were first introduced by [34], and have since had their advantages and drawbacks analyzed extensively in the engineering and control literature [35, 37]. For example, a straightforward metric for describing the efficiency of control from a given driver set is to examine its spectrum $\Lambda(W_T)$. Since the energy cost E_T^* is a quadratic form in W_T^{-1}, we have an immediate upper bound on the control energy:

$$E_T^* \leq \lambda_{\max}(W_T^{-1})\|x_f - A^T x_0\|_2^2 = \lambda_{\min}^{-1}(W_T)\|x_f - A^T x_0\|_2^2 \qquad (3.14)$$

where we have used $\lambda_{\max}(M) \equiv \sup \Lambda(M)$ as the largest eigenvalue of the input matrix and $\lambda_{\min}(M) \equiv \inf \Lambda(M)$ as the smallest. This inequality is derived directly from the Rayleigh quotient. For this reason, a common metric that is used to discriminate between driver sets based on their energy efficiency is the smallest eigenvalue of the Gramian. The minimum eigenvalue of the inverse Gramian matrix (and thus the inverse of the minimum eigenvalue of the Gramian itself) provides an upper bound on the worst-case control energy that may be required to reach some state with a given distance from the final state of the natural autonomous dynamics, with the inequality being strict when $(x_f - A^T x_0)$ is the eigenvector of W_T^{-1} associated with the eigenvalue $\lambda_{\max}(W_T^{-1})$.

Another result of the minimum energy (3.12) taking a quadratic form is that its level curves are nested ellipsoids. For any given control energy budget $\eta \in \mathbb{R}_+$, the states that are reachable within this energy budget fall within the quadric surface whose formula is a level curve of the energy function:

$$(x_f - A^T x_0)W_T^{-1}(x_f - A^T x_0) = \eta \qquad (3.15)$$

This forms an ellipsoid in the state space, centered at the final state $A^T x_0$ of the autonomous dynamics. Following [37], we can denote the ellipsoid of "easily" reachable states (which can be reached with at most 1 unit of control energy) as:

$$\mathcal{E}_{\min}(T) = \{x \in X \mid (x - A^T x_0)W_T^{-1}(x - A^T x_0) \leq 1\} \qquad (3.16)$$

Using the relationship that $\det(W_T^{-1}) = \frac{1}{\det(W_T)}$, we can derive a formula for the volume of this ellipsoid as:

$$V(\mathcal{E}_{\min}) = \frac{\pi^{n/2}}{\Gamma(n/2 + 1)}(\det(W_T))^{1/n} \qquad (3.17)$$

Thus, the determinant of the Gramian gives a natural measure of how many states are potentially reachable with a limited budget, and thus the relative impact of a unit of input energy.

Another commonly used metric for the efficiency of a control schematic is the trace of the inverse Gramian, $\mathrm{Tr}(W_T^{-1})$, which gives a measure of how much energy will be required to control the system to a randomly selected state within a unit sphere surrounding the autonomous final state. Motivated by the relationship $\frac{\mathrm{Tr}(A^{-1})}{n} \geq \frac{n}{\mathrm{Tr}(A)}$, the trace of the Gramian itself $\mathrm{Tr}(W_T)$, may also be used and offers a convenient approximation to the trace of its inverse without the need for potentially expensive matrix inversions.

The trace metrics are particularly interesting because they are not only submodular, but can be optimized in closed form [37]—that is, a control schematic can be easily designed which maximizes this measure. Unfortunately, the control input schematic which maximizes the trace metric may not render the system fully controllable. Submodularity is a desirable feature of efficiency metrics, in general, because optimizations of control input designs is a combinatorial problem. A set function $f : 2^N \to \mathbb{R}$ on the nodes of the network is called *submodular* if it satisfies $f(A) + f(B) \geq f(A \cup B) + f(A \cap B)$ for all subsets $A, B \subseteq N$. Intuitively, submodularity implies decreasing returns; adding a driver to a larger set has a lower impact on a submodular metric than adding a new driver to a smaller set. If a controllability metric of a driver set satisfies submodularity, it provides suboptimality guarantees for a greedy algorithm that progressively adds high-value nodes to the driver set. Notably, while $\det(W_T)$, $\text{Tr}(W_T^{-1})$, and $\text{Tr}(W_T)$ are submodular, spectral measures like $\lambda_{\min}^{-1}(W_T)$ are not necessarily. This means that it is difficult to design driver sets that guarantee reasonable efficiency in the worst-case direction and avoid practically unreachable states [37].

3.5.2 Node-Level Efficiency Metrics

The most recent work in identifying efficient driver sets is [29]. Here, the authors characterize node-level metrics which aggregate well to building an effective and efficient driver set. These measures are based on the "non-normality" of the network system—a phrase used to describe imbalances in the distribution of energy efficiency across the network.

In a directed network system, controlling any node allows a controller to inject energy into the system. This is relatively straightforward given what we have discussed up to this point. For example, the energy injected into the system by only controlling a single node could be summarized by their individual controllability matrix $C_T^{(i)}$, which gives the controllable subspace of the system if *only* node i were controlled. Likewise, we can use this to form the individual controllability Gramian as:

$$W^{(i)} = \sum_{t=1}^{T-1} A^t e_i e_i^\top (A^\top)^t \qquad (3.18)$$

For convenience, we will drop the T subscript from this portion of the discussion, though it should be taken as implicit. Interestingly, for any driver set $V \subseteq N$, the full Gramian is simply the sum of these individual Gramians, so that $W_T^V = \sum_{k \in V} W_T^{(k)}$. By the linearity of the trace, it is clear that the trace of these individual Gramians provide useful information about the capability of a node to inject energy into the network—the trace of a node's individual Gramian is equal to its impact on the trace of the system Gramian. This is referred to by [29] as the *node-to-network* centrality, and is given by:

$$p_i \equiv \text{Tr}(W^{(i)}) \qquad (3.19)$$

This measure alone provides a lot of information about the energy impact of an individual node. It is most useful, however, when combined with another measure called the *network-to-node* centrality. Network-to-node centrality gives a measure of the energy *cost* associated with controlling any node. For example, if a node has no incoming connections, then it is trivial to control. If a node has many incoming connections, on the other hand, then its state may fluctuate dramatically through the course of the dynamics, meaning that a large actuation effort may be required to bring this node to the desired value in each period. This measure is quantified as:

$$q_i \equiv \mathrm{Tr}(M^{(i)})$$ (3.20)

where $M^{(i)}$ is the individual *observability* Gramian. Although we have not discussed observability in this paper, it is a dual problem to controllability. Observability, or state-reconstruction, asks how many (and which) nodes states must be observed in order to perfectly reconstruct the full state vector of the system. As with controllability, it is often the case for network systems that the state vector can be reconstructed reliably from observations of the states of only a limited number of nodes. The optimal state reconstruction, as with the optimal control signal, is related to a Gramian matrix called the observability Gramian. This is given by:

$$M = \sum_{t=1}^{T-1} (A^\top)^t O^\top O A^t$$ (3.21)

where, in this case we have used O as a matrix to describe which states are directly observed. So the observer could see the output vector $y_t = O A^T x_0$. The individual observability Gramians used for computing network-to-node centrality are simply "fictitious" Gramians which would describe the situation in which only node i were observed, or where $O = e_i^\top$, so that:

$$M^{(i)} = \sum_{t=1}^{T-1} (A^\top)^t e_i e_i^\top A^t$$ (3.22)

The trace of this matrix gives, roughly, a measure of how difficult it would be to predict the future state of the node given only information about its current state. Thus, the network-to-node centrality is taken to be:

$$q_i \equiv \mathrm{Tr}(M^{(i)})$$ (3.23)

Intuitively, node-to-network describes how easily a node can control those around it, while node-to-network shows how costly that node itself will be to control directly. Thus, effective input placement will choose driver nodes with high values of p_i and low values of q_i. This can be quantified simply using the difference between the two measures or their ratio. Formally, $r_{\mathrm{diff},i} = p_i - q_i$, and $r_{\mathrm{quot},i} = \frac{p_i}{q_i}$. [29] test these driver selection methods based on these measures in sample networks, and find that they can be used to reliably produce efficient control schematics. As with the controllability metrics above, we computed these efficiency statistics for the 108th

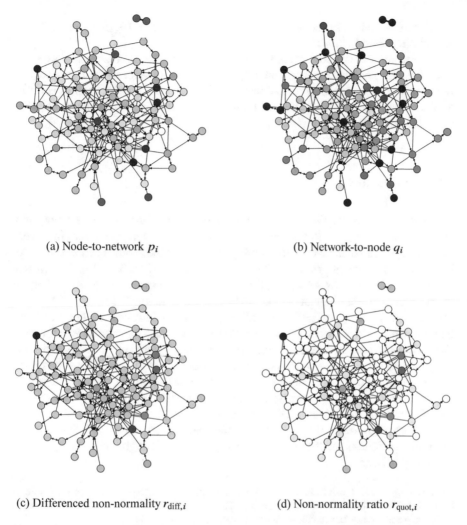

(a) Node-to-network p_i (b) Network-to-node q_i

(c) Differenced non-normality $r_{\text{diff},i}$ (d) Non-normality ratio $r_{\text{quot},i}$

Fig. 3.4 Node-level control efficiency metrics for the 108th Senate of the United States of America (2003–2005). **a** Node-to-network p_i. **b** Network-to-node q_i. **c** Differenced non-normality $r_{\text{diff},i}$. **d** Non-normality ratio $r_{\text{quot},i}$. In **a**, p_i is plotted on a log scale, with darker colors representing higher values of node-to-network centrality. In (2), q_i is also colored on a log scale and darker colors represent nodes that are more costly to control due to their network-to-node centrality. While root nodes are highly influential, they can also be expensive to control, depending on their self-weights. Non-normality must be balanced—if there are nodes that influence far more than they are influenced, it must mean that there are also nodes who influence less than they are influenced. This means that $r_{\text{diff},i}$ must always sum to 0; as long as there are nodes with positive values of $r_{\text{diff},i}$, there must also be nodes with negative values. In **c** and **d**, $r_{\text{diff},i}$ and $r_{\text{quot},i}$ are shown, with more efficient control nodes as darker colors. Summary statistics for these measures can be found in Table 3.1

US Senate. Results of these computations are visualized in Fig. 3.4, and sufficient statistics reported in Table 3.1.

For the purposes of corruption research, the crucial question is exactly *which* of these (or other) control measures are important to the corrupt nodes in the network, which measures can be manipulated by the nodes, and thus which measures could be used to identify certain nodes that are high-risk for exhibiting manipulative behavior in exchange for monetary payment.

3.6 Discussion

The link between corruption and the controllability of social networks is a new field, and there is a need for more research into the specific mechanisms that link these two concepts. In particular, controllability metrics for more advanced social learning dynamics—such as Friedkin-Johnsen [10], DeGroot-Friedkin [20], or Bayesian models [1, 12] could prove a fruitful area of future research, especially in light of the substantial energy impact of stubbornness and self-loops. In addition, while new insights into the design of efficient control systems could prove useful in the identification of high-risk areas in a network system, there is also a growing need for more theoretical and empirical research into the connections between these concepts and the particular role of opinion dynamics in state capture. For the purposes of curtailing corruption, it is crucial to gain a better understanding of both the incentive structure and institutional features that support corruption in these environments, and which specific mechanisms for control they enable. We hope that this review will provide a useful foundation for future research into the specific mechanics underlying "quid-pro-quo" corruption and the spread of misinformation through legislative, political, and social networks.

Acknowledgements I would like to thank participants at the corruption networks satellite of NetSci2020 for helpful comments and advice. Thanks also to the Charles & Persis Rockwood Fellowship and the L. Charles Hilton Center for their generous support.

References

1. Acemoglu D, Dahleh MA, Lobel I, Ozdaglar A (2011) Bayesian learning in social networks. Rev Econ Stud 78(4):1201–1236
2. Acemoglu D, Ozdaglar A, ParandehGheibi A (2010) Spread of (mis) information in social networks. Games Econom Behav 70(2):194–227
3. Battaglini M, Patacchini E, Rainone E (2019) Endogenous social connections in legislatures. Technical report, National Bureau of Economic Research
4. Bindel D, Kleinberg J, Oren S (2015) How bad is forming your own opinion? Games Econom Behav 92:248–265
5. Briatte F (2016) Network patterns of legislative collaboration in twenty parliaments. Netw Sci 4(2):266–271

6. Chandrasekhar A (2016) Econometrics of network formation. The Oxford handbook of the economics of networks, pp 303–357
7. da Cunha BR, Gonçalves S (2018) Topology, robustness, and structural controllability of the brazilian federal police criminal intelligence network. Appl Netw Sci 3(1):1–20
8. DeGroot MH (1974) Reaching a consensus. J Am Stat Assoc 69(345):118–121
9. Dion JM, Commault C, Van der Woude J (2003) Generic properties and control of linear structured systems: a survey. Automatica 39(7):1125–1144
10. Friedkin NE, Johnsen EC (1990) Social influence and opinions. J Math Sociol 15(3–4):193–206
11. Gaitonde J, Kleinberg J, Tardos E (2020) Adversarial perturbations of opinion dynamics in networks. arXiv preprint arXiv:2003.07010
12. Gale D, Kariv S (2003) Bayesian learning in social networks. Games Econom Behav 45(2):329–346
13. Golub B, Jackson MO (2010) Naive learning in social networks and the wisdom of crowds. Am Econ J: Microecon 2(1):112–49
14. Golub B, Jackson MO (2012) How homophily affects the speed of learning and best-response dynamics. Q J Econ 127(3):1287–1338
15. Granovetter M (2007) The social construction of corruption. On capitalism 15
16. Gross JH, Kirkland JH, Shalizi C (2012) Cosponsorship in the us senate: a multilevel two-mode approach to detecting subtle social predictors of legislative support. Unpublished manuscript
17. Hautus ML (1969) Controllability and observability conditions of linear autonomous systems. Indagationes Mathematicae (Proc) 72:443–448
18. Huckfeldt R, Pietryka MT, Reilly J (2014) Noise, bias, and expertise in political communication networks. Soc Netw 36:110–121
19. Huckfeldt R, Sprague J (1987) Networks in context: the social flow of political information. Am Polit Sci Rev 81(4):1197–1216
20. Jia P, MirTabatabaei A, Friedkin NE, Bullo F (2015) Opinion dynamics and the evolution of social power in influence networks. SIAM Rev 57(3):367–397
21. Jia T, Barabási AL (2013) Control capacity and a random sampling method in exploring controllability of complex networks. Sci Rep 3:2354
22. Jia T, Liu YY, Csóka E, Pósfai M, Slotine JJ, Barabási AL (2013) Emergence of bimodality in controlling complex networks. Nat Commun 4:2002
23. Kalman RE (1963) Mathematical description of linear dynamical systems. J Soc Ind Appl Math, Ser A: Control 1(2):152–192
24. Korte B, Vygen J, Korte B, Vygen J (2012) Combinatorial optimization, vol 2. Springer, Berlin
25. Lazer D, Rubineau B, Chetkovich C, Katz N, Neblo M (2010) The coevolution of networks and political attitudes. Polit Commun 27(3):248–274
26. Li G, Hu W, Xiao G, Deng L, Tang P, Pei J, Shi L (2015) Minimum-cost control of complex networks. New J Phys 18(1):013,012 (2015)
27. Lin CT (1974) Structural controllability. IEEE Trans Autom Control 19(3):201–208
28. Lindmark G, Altafini C (2018) Minimum energy control for complex networks. Sci Rep 8(1):3188
29. Lindmark G, Altafini C (2020) Centrality measures and the role of non-normality for network control energy reduction. IEEE Control Syst Lett 5(3):1013–1018
30. Liu YY, Slotine JJ, Barabási AL (2011) Controllability of complex networks. Nature 473(7346):167
31. Liu YY, Slotine JJ, Barabási AL (2012) Control centrality and hierarchical structure in complex networks. Plos one 7(9):e44,459
32. Liu Z, Ma J, Zeng Y, Yang L, Huang Q, Wu H (2014) On the control of opinion dynamics in social networks. Physica A 409:183–198
33. Moosavi-Dezfooli SM, Fawzi A, Fawzi O, Frossard P (2017) Universal adversarial perturbations. In: Proceedings of the IEEE conference on computer vision and pattern recognition, pp 1765–1773
34. Müller P, Weber H (1972) Analysis and optimization of certain qualities of controllability and observability for linear dynamical systems. Automatica 8(3):237–246

35. Pasqualetti F, Zampieri S, Bullo F (2014) Controllability metrics, limitations and algorithms for complex networks. IEEE Trans Control Netw Syst 1(1):40–52
36. Solimine PC (2020) Political corruption and the congestion of controllability in social networks. Appl Netw Sci 5:1–20
37. Summers TH, Cortesi FL, Lygeros J (2015) On submodularity and controllability in complex dynamical networks. IEEE Trans Control Netw Syst 3(1):91–101
38. Wang WX, Ni X, Lai YC, Grebogi C (2012) Optimizing controllability of complex networks by minimum structural perturbations. Phys Rev E 85(2):026,115
39. Yan G, Tsekenis G, Barzel B, Slotine JJ, Liu YY, Barabási AL (2015) Spectrum of controlling and observing complex networks. Nat Phys 11(9):779–786
40. Yu X, Liang Y, Wang X, Jia T (2021) The network asymmetry caused by the degree correlation and its effect on the bimodality in control. Phys A: Stat Mech Appl 125868
41. Yuan Z, Zhao C, Di Z, Wang WX, Lai YC (2013) Exact controllability of complex networks. Nat Commun 4:2447

Chapter 4
Predicting Corruption Convictions Among Brazilian Representatives Through a Voting-History Based Network

Tiago Colliri and Liang Zhao

Abstract While analyzing voting data concerning almost 30 years of legislative work from Brazilian representatives, we have noticed the formation of some sort of "corruption neighborhoods" in the resulting congresspeople network, indicating the possible existence of an (until then) unsuspected relationship between voting history and convictions of corruption or other financial crimes among legislators. This finding has motivated us to develop a predictive model for assessing the chances of a representative for being convicted of corruption or other financial crimes in the future, solely based on how similar are his past votes and the voting record of already convicted congresspeople. In this study, we present the main results obtained from this investigation.

4.1 Introduction

Corruption affects the society negatively in several ways, from holding back businesses [1], to the waste of public spending and investment [2] and the weakening of democratic systems [3, 4]. Predicting the incidence of corruption, specially at the individual level, is a challenging task, because criminals constantly develop increasingly advanced mechanisms to cover their infractions. More recently, governments around the world have been taking measures to increase their transparency by making public administration data accessible to the population. This phenomenon had encouraged researchers and members of the society to develop new mechanisms to monitor and analyze such data, through multidisciplinary approaches, thus contributing to increase the levels of accountability in the public sector [5].

Network-based techniques have already proven successful in the analysis of politics-related data, such as in the legislators' relations through bill co-sponsorship

T. Colliri (✉)
Institute of Mathematics and Computer Science, University of Sao Paulo, Sao Carlos, Brazil
e-mail: tcolliri@usp.br

L. Zhao
Faculty of Philosophy, Science, and Letters, University of Sao Paulo, Ribeirao Preto, Brazil
e-mail: zhao@usp.br

© The Author(s), under exclusive license to Springer Nature Switzerland AG 2021
O. M. Granados and J. R. Nicolás-Carlock (eds.), *Corruption Networks*,
Understanding Complex Systems, https://doi.org/10.1007/978-3-030-81484-7_4

data [6, 7] and through roll-call voting data [8–12]. Additionally, this type of approach
has also been applied on the analysis of crimes-related data, such as in the correlation between social capital and the risk of corruption in local governments contracts
[13] and in the identification of missing links among the members of an Italian mafia
group [14]. Another interesting work, still related to this type of application, used link
prediction techniques to uncover hidden relations among Brazilian politicians cited
on corruption scandals [15]. In a more recent work, a network-based approach has
been applied for modeling the dynamics of a major corruption scandal occurred in
Mexico involving embezzlement activities, contributing to provide systematic evidence on which corporate characteristics are likely to signal corruption in public
procurement [16].

The term *complex network* refers to a large scale *graph* with non-trivial connection
patterns [17]. Some examples of complex networks in the real world include the
internet [18], biological neural networks [19], food chains [20], blood distribution
networks [21] and power grid distribution networks [22]. Additionally, there are also
various network-based models designed to perform *machine learning* related tasks,
such as *clustering* [23–25], *classification* [26–29] and *regression* [30, 31]. More
recently, there was the introduction of *temporal networks*, which allows the inclusion
of the *time* dimension in the study of graphs. Examples of temporal networks that
can be found in the real world include social networks, one-to-many information
dissemination (e.g., emails or blogs), cell-biology networks, brain networks, traffic
networks, and mobile communication networks [32].

In this work, we start by using a network-based approach to build a temporal
network from voting data regarding almost 30 years of legislative work in the Brazilian House of Representatives. Each node in this network corresponds to a different
legislator, who has voted at least once in the House during that period, and the edges
between each pair of nodes are created according to the voting history similarity
between them. Afterwards, we investigate whether this built network can be used
to predict cases of conviction for corruption or other financial crimes among the
congresspeople, based on the emerging topological structure formed by convicted
congresspersons and their neighbors in the network. The dataset used in the application has been created especially for this work, comprising the votes of a total of
2,455 congresspeople and 3,407 legislative voting sessions, from 1991 until 2019.
The results obtained from this study were originally published as part of the analyses
made in [33] and, for this revised version, we have updated the figures, as well as
inserted a new one, as Fig. 4.4 in Sect. 4.3, to display the legislators network used in
the corruption prediction task. Additionally, we have also added Tables 4.1 and 4.2,
in Sect. 4.2, to help demonstrate how the legislators temporal network is built, by
using a simple voting dataset as example.

Table 4.1 Illustration showing voting sessions outcome and respective resulting legislators' temporal network slices, based on the voting history similarity

	Votes		
Voting Session	**1**	**2**	**3**
Legislator 0	No	Obstruction	Yes
Legislator 1	No	Obstruction	Yes
Legislator 2	No	Yes	Yes
Legislator 3	Yes	-	-
Legislator 4	Yes	No	Yes
Legislator 5	-	No	No
Resulting Network			

Table 4.2 Final values of weight matrix $W^{(t)}$ in the example voting dataset ($t = 3$). The values in the main diagonal are changed to 0 for avoiding self-loops in the built network

	0	1	2	3	4	5
0	0	3	1	−1	−3	0
1	3	0	1	−1	−3	0
2	1	1	0	−1	−3	0
3	−1	−1	−1	0	1	0
4	−3	−3	−3	1	0	0
5	0	0	0	0	0	0

4.2 Methodology

The methodology used in this study is described below. All network analyses performed in this study were implemented by using the igraph [34] and Teneto [35] Python packages.

4.2.1 Database

The data for our analysis are collected from the official website of the Brazilian House of Representatives [36]. The obtained datasets comprise the outcome of all voting sessions of legislative bills deliberated in the House, from May 22, 1991 until Feb 14, 2019. As preprocessing, we have performed a thorough data cleansing in the database to detect and fix possible errors, such as duplicated names among legislators and also misprints. Each session comprises the following information: the bill to be considered, the voting date, and the representatives who have attended the session and voted. Additionally, the following information is provided for each voter:

- ID (a unique number for each congressperson),
- Full Name,
- Political Party, and
- Vote.

The voting system in the Brazilian House consists of four types of votes:

- *Yes*: if the representative approves the bill;
- *No*: if the representative disapproves the bill;
- *Abstention*: if the representative deliberately chooses to not take part in the voting;
- *Obstruction*: analogous to abstention, except that abstention counts for *quorum* effects, while obstruction does not count for it.

   The final database contains the voting outcome from a total of 1,656,547 votes from 3,407 sessions, and 2,455 different congresspersons. To perform our analyses, we map each of the voting sessions to a static network, where each node represents a legislator who attended that session and the edges between them are created according to their respective voting history similarity, pairwise. The Brazilian House currently has 513 seats, hence this is the maximum number of nodes in the legislators network resulting from each voting session.

   Additionally, we also made an extensive research in several media vehicles to collect data concerning which congresspersons in the database have already been officially convicted for corruption or other financial crimes, such as: embezzlement, improbity, misappropriation of public funds, money laundering, peculation, or crimes against the Public Administration. This task resulted in the identification of 21 legislators who have been convicted for corruption and 12 legislators who have been convicted for other financial crimes, out of the total 2,455 congresspeople in the database. This additional information was confirmed from legitimate Brazilian judiciary sources as well, such as the Federal Supreme Court (Supremo Tribunal Federal) [37].

4.2.2 Legislators Temporal Network Generation

A network can be characterized as graph $G = (\mathcal{V}, \mathcal{E})$, where $\mathcal{V}$ is a set of nodes and $\mathcal{E}$ is a set of tuples representing the edges between each pair of nodes $(i, j) : i, j \in \mathcal{V}$. The edges in $\mathcal{E}$ are usually provided in the form of a square matrix M, with size equal to the number of nodes in the network and binary values, in case of unweighted graphs. In order to convert a list of static networks into a temporal network $\mathcal{G} = \{G^{(t=0)}, G^{(t=1)}, G^{(t=2)}, \ldots, G^{(t=n)}\}$, where t is the time step, one then needs to insert a new dimension $\mathcal{D}$ in the formal network definition, such that it becomes $\mathcal{G} =$

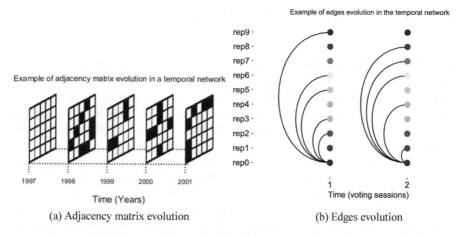

Fig. 4.1 **a** Example of the adjacency matrix evolution in a temporal network whose dimension $\mathcal{D}$ is measured in terms of years. **b** Edges evolution. Illustration showing the graphlets evolution in time (in this work measured in terms of voting sessions). When $t = 1$, node 0 is connected to nodes 2, 3, 5, 6 and 9. Following, when $t = 2$, it disconnects from nodes 2, 3, 5 and 9 and becomes connected to nodes 1, 4, 7 and 8

$(\mathcal{V}, \mathcal{E}, \mathcal{D})$, where $\mathcal{D}$ represents the temporal network slices. This process is illustrated in Fig. 4.1. One way of achieving this is by generating a matrix for representing the edges in $\mathcal{E}$ as a *triplet* $(i, j, t) : i, j \in \mathcal{V}, t \in \mathcal{D}$, also known as dynamic *graphlets* [38] (Fig. 4.1b). The final result from this conversion process is a *multilayer network*, with each layer representing a static temporal slice of a single main graph (Fig. 4.1a) and, for cases when the dimension $\mathcal{D}$ is a set of indices ordered by time, one can also refer to this graph as a *temporal network* [35].

In this work, we start by converting each voting session in the database into a static network, based on their "date" attribute, in ascending order. For each session, a square votes matrix $M^{(t)}$, of size $d \times d$, is created, where t is the session's sequential number, or time step, and d is the total number of legislators who attended that session. Hence, as mentioned earlier, the maximum possible value for d is 513, because this is the total number of seats in the Brazilian House. Each element $M_{ij}^{(t)}$ has a binary value, assuming:

$$M_{ij}^{(t)} = \begin{cases} 1, & \text{if congresspersons } i \text{ and } j \text{ voted alike in session } t, \\ -1, & \text{otherwise.} \end{cases} \tag{4.1}$$

The values of each voting matrix $M^{(t)}$ are accumulated in a distinct weight matrix $W^{(t)}$, whose size is equal to the total number of legislators who have registered at least one vote in the House, until session t. Therefore, each element $W_{ij}^{(t)}$ from this matrix returns the accumulated weight between congresspersons i and j, or the *voting history similarity* between them, until session t. Formally, the current value of each element $W_{ij}^{(t)}$ is yielded by:

$$W_{ij}^{(t)} = \sum_t M_{ij}^{(t)}. \tag{4.2}$$

Note that, from Eq. 4.2, the value of $W_{ij}^{(t)}$ will range from $-t$, in case congressperson i have always voted different from the representative j, until $+t$, when congressperson i have always voted similarly to the representative j, until session t. In this study, we are assuming that, in the former case, legislators i and j have complete opposite political views, while, in the latter case, legislators i and j are very politically aligned, up to the instant t.

The static slices $G^{(t)}$, for each voting session t, which will form the legislators temporal network G, are obtained from the values in weight matrix $W^{(t)}$. For this end, we opt to connect each legislator to the one(s) with the highest weight(s) associated to him in $W^{(t)}$, i.e., each legislator will be connected to the representative(s) who are most politically aligned to him, in terms of their voting history similarity, up to the instant t. Thus, the edges between each pair of nodes representing congresspersons i and j, in each temporal network slice $G^{(t)}$, are yielded by:

$$G_{ij}^{(t)} = \begin{cases} 1, & \text{if } W_{ij}^{(t)} = \max_{\forall x \in W_i^{(t)}} x \\ 0, & \text{otherwise .} \end{cases} \tag{4.3}$$

Note that, from Eq. 4.3, most of the nodes in $G^{(t)}$ will have only one outbound edge, connecting it to the node most politically aligned to it. The only exceptions for this rule are the situations when $\max_{\forall x \in W_i^{(t)}} x$ returns more than one element, which in this case will result in two or more outbound edges originated from node i.

One can also consider, in this step of the technique, the possibility of binning the legislators voting similarities by predetermined time slices, such as per year or per presidential term, for instance. In our case, we have performed some preliminary tests considering this possibility, and the results from these tests indicated that, for this specific database, it is necessary to process a large number of voting sessions before a clear topological pattern emerges in the temporal network. A possible reason for this may be in the fact that most representatives serve for only 4 years, which is the term length in Brazil, and only few of them get reelected. Additionally, we noted many cases of officially elected congresspeople leaving their seats for interim successors, to run for higher political positions, such as governors or senators. This specific feature in our database also contributes to prevent the formation of a clear voting similarity pattern in the legislators temporal network's topology, in the short-term and medium-term. Therefore, we have opted for taking the legislators' complete voting history into consideration for generating the edges in the network.

In order to illustrate how the temporal network G is generated, let us now consider a simple dataset, comprising only 3 voting sessions, each one attended by 5 legislators (Table 4.1). In voting session 1, legislators 0, 1 and 2 voted "No", while legislators 3 and 4 voted "Yes". This outcome results in a network with two components, with all nodes being connected with equal weights (+1, in this case). In session 2, legislator

2 votes different from legislators 0 and 1, and legislator 3 is replaced by legislator 5, which votes similar to legislator 4. Now, the temporal network is updated, with legislator 2 isolating from the others (because his current voting similarity score is 0), and a new node is inserted to represent legislator 5, connected to legislator 4. Note that the node from legislator 3 still remains in the temporal network, in the same position it was before, since its voting similarity score remains unchanged. In the third voting session, legislators 0, 1 and 2 again vote alike, so legislator 2 reconnects to legislators 0 and 1 in the network, with a similarity score of +1 with both of them. Legislators 0 and 1 reach a maximum similarity score of +3 with each other, since they voted alike in all three sessions, so they are reciprocally connected. Legislator 5 is the only one who votes "No" in this session, which results in his isolation from the others since his voting similarity score now is 0. The final values of weight matrix $W^{(t)}$, i.e., when $n = 3$ in the example, are shown in Table 4.2.

It is important to stress that, as it is shown in the example from Tables 4.1 and 4.2, when a new congressperson is inserted in the network – as in the case of new legislative elections or because of a resignation, for instance – he or she does not inherit any voting data from the congressperson who previously occupied its node in the network (or "seat" in the House, so to speak). For such cases, a new row and a new column is inserted in weight matrix $W^{(t)}$ to record the voting similarity between the new node, representing the new congressperson, and all the other elements in $W^{(t)}$. Another important aspect to be emphasized in the legislators' temporal network building process is that we opt to not take into account the attribute "political party" from the legislators when generating the edges. We have decided to proceed this way because our goal, in this study, is to capture the political affinities among congresspeople beyond their party affiliations, solely by considering their voting records. This decision can be justified inasmuch as there are currently 33 political parties officially in Brazil, and this exaggerated number of parties eventually results in a consistent attenuation of the ideological differences between them.

4.2.3   Corruption Prediction Model

Let us now proceed to describe how we assess whether a representative is more or less likely to be convicted of corruption or other financial crimes in the future by analyzing the built voting-history-based legislators network. To accomplish this task, we make use of link prediction techniques applied on a subgraph of the congresspeople network, formed from a subset of the nodes comprising only already convicted representatives and their respective neighbors, i.e., the congresspersons identified as the most politically aligned to them, according to our model. The prediction tests are performed in a supervised learning manner, in which we take the top n links predicted by each considered technique whose source node is a convicted congressperson, and label their target nodes as being convicted ones as well. Following, we provide more details regarding the prediction model.

The final weight matrix $W^{(t)}$ from the built legislators network has a total of 2,455 nodes, representing all congresspeople in our database who have voted at least once in the House, from 1991 until 2019. While browsing this final resulting network, we noticed that, oftentimes, the neighbor of a convicted congressperson was a convicted one as well, seemingly forming some sort of "corruption neighborhoods" in the built network, so to speak. This unexpected emerging pattern motivated us to investigate this aspect further, by applying algorithms which consider the network topological structure for predicting missing links. Below, we list the required steps for the corruption prediction model evaluated in this study.

1. Create a subgraph from the built legislators network resulting from matrix $W^{(t)}$, comprising only convicted congresspeople and their respective neighbors, from both incoming and outgoing edges.
2. Convert the network resulting from the subgraph to an *undirected* one, and delete all existing links between convicted labeled nodes.
3. Apply the link prediction technique on the network.
4. Take the top n link predictions whose source node is convicted and label their target nodes as convicted ones as well.

The accuracy achieved by each link prediction technique is evaluated in a supervised learning manner, by inspecting how many of its target nodes are indeed labeled as convicted ones in the database. Please note that, by making use of link prediction techniques in this task, we are thus considering the legislators network topological structure for conviction prediction purposes.

The link prediction problem is often studied in social networks, and can be formulated as "to what extent can one predict which links will occur in a network based on older or partial network data" [39]. This task can be categorized into two different types: *missing link prediction*, which involves predicting links based on an incomplete or damaged version of a network, and *future link prediction*, which involves predicting links in a future snapshot of the network based on its current state [40]. The predictors, by their turn, can be categorized into two different groups, as listed below.

- *Local predictors*: based on the neighborhoods of the source and target nodes (e.g., Cosine [41, 42], Maximum and Minimum Overlap [43], NMeasure [44] and Pearson [45]).
- *Global predictors*: based on measures that evaluate how the source and target nodes may be related in the network, even if they do not share any common neighbors (e.g., Katz [46] and Rooted PageRank [47]).

In this study, we test the following 6 link prediction techniques in the corruption prediction task: Cosine, MinOverlap, NMeasure, Pearson, Rooted PageRank and Random (for comparison purposes). All prediction techniques are implemented by using the tool introduced in [40], with default parameter values.

4.3   Results

We start this section by presenting an example of one of the networks resulting from our analysis, in Fig. 4.2. This network is built from data of the voting session for bill PEC 77/2003, occurred on September 19, 2017, which was attended by all 513 Brazilian congresspeople. The outbound edges connect each congressperson to the one most politically aligned to him, based on their voting history similarity. Therefore we can say that the hubs in the network, within this context, are congresspersons who represent a "local majority" in terms of voting, or the majority within a local neighborhood. We have also applied a community detection technique in this figure, by using the fast greedy algorithm [48], through the tool introduced in [34], for the sake of highlighting the emerging "voting aggregation patterns" among congresspeople in the Brazilian House of Representatives. These voting patterns likely occur due to aggregation factors among the legislators, such as political parties, civic associations or interest groups.

In Fig. 4.3, we show the final resulting network from our analysis, originated from the final matrix $W^{(t)}$, with $t = 3,407$, i.e., the total number of voting sessions in the database. This graph comprises all 2,455 representatives in the database, who have voted at least once in the Brazilian House, from May 22, 1991 until February 14, 2019. The red nodes indicate the 33 congresspeople currently identified as convicted ones, in our research. While browsing the nodes of this network, we have noted that the neighbors of a convicted congressperson were oftentimes convicted ones as well. This unexpected emerging feature has motivated us to further investigate whether the nodes of convicted legislators indeed tend to stay close to each other in the network, thus forming some sort of "corruption neighborhoods", so to speak. For this end, we build a separated network, from a subgraph of this final network, containing only the nodes from convicted representatives, along with their respective neighbors, from both incoming and outgoing edges. It is worth noting that these neighbors may be convicted ones as well or not. As preprocessing, before applying the link prediction techniques, we convert this network to an *undirected* one and remove all existing links between two nodes labeled as convicted from the network (5 edges in total). The subgraph resulting from this preprocessing is shown in Fig. 4.4. It comprises 211 legislators, with 33 of them being convicted, and 1,374 edges.

In order to predict new conviction cases among representatives, we apply link prediction techniques in the network shown in Fig. 4.4. A total of 5 different link prediction techniques plus a Random method (for comparison purposes) were considered for this task. We take the top 10 links predicted by each technique, having convicted nodes as a source, and labeled their target nodes as being convicted ones as well. The accuracy of the model is assessed by inspecting how many of these predictions are correct, according to the convictions information in the database.

In Fig. 4.5, we display the accuracy achieved by each considered technique in the conviction prediction task. These results indicate that it is possible to achieve a high performance, in terms of corruption prediction, when considering the topological structure of the built legislators network. The Cosine, NMeasure and Pearson tech-

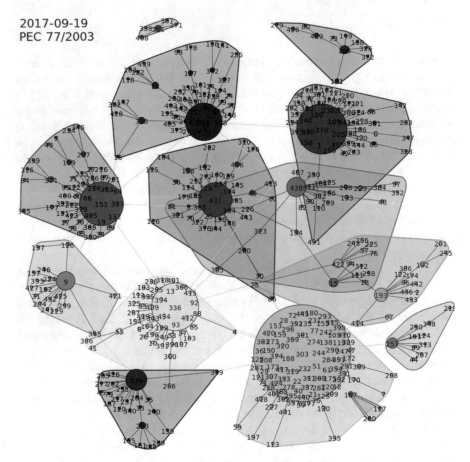

Fig. 4.2 Representation of one of the temporal network slices generated in our analysis, from data regarding voting session of bill PEC 77/2003, occurred on 2017-09-19. Each node represents one of the 513 congresspersons who have voted in this session, and the edges are created based on their voting history similarity. The communities were detected through the fast greedy algorithm. This graph was converted to an undirected one only for visibility purposes

niques were able to achieve an accuracy of 0.9 in this task, i.e., 9 out of the 10 links predicted by these techniques turned out to be correct, with their target nodes being convicted congresspersons. Rooted PageRank and MinOverlap techniques have presented slightly lower performances in this task, with an accuracy of 0.7 and 0.5, respectively. Note that the Random technique was not able to predict not even one link correctly in this task, which indicates both a higher level of difficulty involved in it and also that prediction techniques based on the graph topological structure are considerably more appropriate for this end.

To better understand the reason why some link prediction techniques have performed better than others in the corruption prediction task, we show, in Fig. 4.6, a

Fig. 4.3 Representation of the final network resulting from our analysis, with all 2,455 representatives in the database, who voted at least once in the Brazilian House of Representatives, from May 22, 1991 until Feb 14, 2019. The edges are created according to their voting record similarity. Convicted representatives are denoted by the red color (33 in total)

comparison between the top 10 links predicted by Cosine and Rooted PageRank techniques. Note that the target nodes from the links predicted by Cosine technique are all nearer their respective source nodes while, on the other hand, the target nodes from the links predicted by Rooted PageRank technique are often more far from their respective source nodes. The explanation for this difference is in the fact that techniques such as Pearson, Cosine and NMeasure are all *local predictors*, i.e., they take into account the neighborhood from each pair of nodes for assessing the possibility of existing a hidden link between them in the network. Thus, the fact that such type of predictors have performed better in the corruption prediction task is a strong sign that our initial hypothesis regarding the seemingly existence of "corruption neighborhoods" in the legislators network is correct, pointing in the direction that there is indeed a correlation between voting history and convictions of corruption or other financial crimes among Brazilian congresspeople.

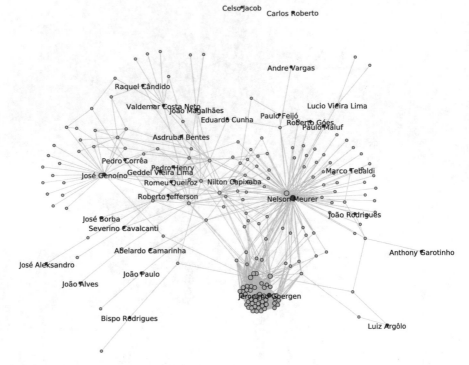

Fig. 4.4 Subgraph of the consolidated network, shown in Fig. 4.3, with only the 33 convicted representatives (in red) and their respective neighbors. The network is converted to an undirected one and links between two convicted congresspeople are suppressed (5 links in total), before applying the link prediction techniques. We opted for displaying only the names of legislators who were identified as officially convicted in this figure

4.4   Final Remarks

In this work, we introduce a technique to generate a temporal network from legislative voting data, and apply it on a dataset specially built for this study, comprising almost 30 years of votes from Brazilian congresspeople. We also investigate the possibility of using the resulting network for predicting corruption-conviction cases among legislators, by using link prediction techniques that consider the topological structure emerging from convicted congresspersons and their neighbors in the network. The obtained results in this task are encouraging, especially the ones achieved by techniques based on local predictors, i.e., based on the neighborhoods of the source and target nodes, which is an indication that there is indeed a correlation between voting history similarity and corruption convictions among Brazilian representatives.

Although in this study we have focused our analyses on aspects regarding corruption and other financial crimes among legislators, it is important to highlight that the congresspeople network resulting from the voting history similarity between them

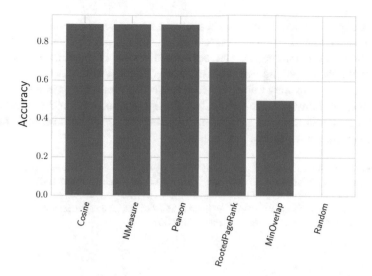

Fig. 4.5 Accuracy achieved by each link prediction technique in the conviction prediction task

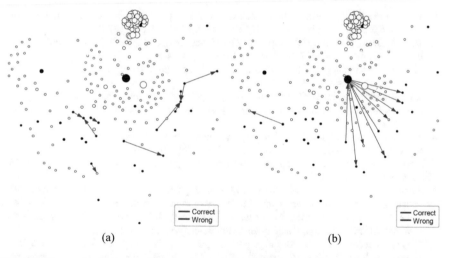

(a)   (b)

Fig. 4.6 Comparison between the link prediction outputs from **a** Cosine and **b** Rooted PageRank techniques. In **a** Links predicted by Cosine (9 correct, 1 wrong). In **b** Links predicted by Rooted PageRank (7 correct, 3 wrong). The black color denotes convicted legislators. A link prediction is considered correct when its target node is also labeled as convicted in the database. All other links are removed from the network for visibility purposes

may also be employed in other types of analyses, focused on the political aspect of the connections. Within this context, one can further explore, for instance, the emerging aggregations verified in the temporal network slices to perform analyses comparing the level of cohesive voting behavior between political parties, and how the changes in these aggregations may impact the legislative decision-making in the House. Or, as another possibility, one can also analyze the aggregation patterns in these network slices for evaluating and mapping the similarities and differences between political parties, in terms of legislative voting behavior.

Acknowledgements This work is supported in part by the Sao Paulo State Research Foundation (FAPESP) under grant numbers 2015/50122-0, the C4AI under grant number FAPESP/IBM/USP: 2019/07665-4, the Brazilian National Council for Scientific and Technological Development (CNPq) under grant number 303199/2019-9 and the Coordenação de Aperfeiçoamento de Pessoal de Nível Superior - Brasil (CAPES) - Finance Code 001.

References

1. Wilhelm PG (2002) International validation of the corruption perceptions index: implications for business ethics and entrepreneurship education. J Bus Ethics 35(3):177–189
2. Tanzi V, Davoodi H (1998) Corruption, public investment, and growth. In: The Welfare State, Public Investment, and Growth. Springer, Berlin, pp 41–60
3. Linde J, Erlingsson GÓ (2013) The eroding effect of corruption on system support in s weden. Governance 26(4):585–603
4. Rose-Ackerman S (2013) Corruption: a study in political economy. Academic, New York
5. Hale TN (2008) Transparency, accountability, and global governance. In: Global governance, pp 73–94
6. Kirkland JH, Gross JH (2014) Measurement and theory in legislative networks: the evolving topology of congressional collaboration. Soc Netw 36(1):97–109
7. Neal ZP (2018) A sign of the times? Weak and strong polarization in the US Congress, 1973–2016. Social Networks
8. Andris C, Lee D, Hamilton MJ, Martino M, Gunning CE, Selden JA (2015) The rise of partisanship and super-cooperators in the U.S. House of representatives. PLoS ONE 10(4):1–14
9. Dal Maso C, Pompa G, Puliga M, Riotta G, Chessa A (2014) Voting behavior, coalitions and government strength through a complex network analysis. PLoS One 9:12
10. Moody J, Mucha PJ (2013) Portrait of political party polarization. Netw Sci 1(1):119–121
11. Waugh AS, Pei L, Fowler JH, Mucha PJ, Porter MA (2009) Party polarization in congress: a network science approach. arXiv preprint arXiv:0907.3509
12. Victor JN, Montgomery AH, Lubell M (2017) The Oxford handbook of political networks. Oxford University Press, Oxford
13. Wachs J, Yasseri T, Lengyel B, Kertész J (2019) Social capital predicts corruption risk in towns. R Soc Open Sci 6(4):182103
14. Berlusconi G, Calderoni F, Parolini N, Verani M, Piccardi C (2016) Link prediction in criminal networks: a tool for criminal intelligence analysis. PLoS One 11:4
15. Ribeiro HV, Alves LG, Martins AF, Lenzi EK, Perc M (2018) The dynamical structure of political corruption networks. J Complex Netw 6(6):989–1003
16. Luna-Pla I, Nicolás-Carlock JR (2020) Corruption and complexity: a scientific framework for the analysis of corruption networks. Appl Netw Sci 5(1):1–18
17. Albert R, Barabási AL (2002) Statistical mechanics of complex networks. Rev Mod Phys 74:47–97

18. Faloutsos M, Faloutsos P, Faloutsos C (1999) On power-law relationships of the internet topology. ACM SIGCOMM Comput Commun Rev 29(4)
19. Sporns O (2002) Network analysis, complexity, and brain function. Complexity 8(1):56–60
20. Montoya JM, Solé RV (2002) Small world patterns in food webs. J Theor Biol 214(3):405–412
21. West GB, Brown JH, Enquist BJ (2009) A general model for the structure, and allometry of plant vascular systems. Nature 400:125–126
22. Albert R, Albert I, Nakarado GL (2004) Structural vulnerability of the north American power grid. Phys Rev 69(2):025103
23. Liu W, Suzumura T, Ji H, Hu G (2018) Finding overlapping communities in multilayer networks. PLOS One 13(4):e0188747
24. Palla G, Derényi I, Farkas I, Vicsek T (2005) Uncovering the overlapping community structure of complex networks in nature and society. Nature 435(7043):814–818
25. Silva TC, Zhao L (2012) Stochastic competitive learning in complex networks. IEEE Trans Neural Netw Learn Syst 23(3):385–398
26. Silva TC, Zhao L (2012) Network-based high level data classification. IEEE Trans Neural Netw Learn Syst 23(6):954–970
27. Colliri T, Ji D, Pan H, Zhao L (2018) A network-based high level data classification technique. In: 2018 international joint conference on neural networks (IJCNN). IEEE, pp 1–8
28. Carneiro MG, Zhao L (2017) Organizational data classification based on the importance concept of complex networks. IEEE Trans Neural Netw Learn Syst 29(8):3361–3373
29. Backes AR, Casanova D, Bruno OM (2013) Texture analysis and classification: a complex network-based approach. Inf Sci 219:168–180
30. Loglisci C, Malerba D (2017) Leveraging temporal autocorrelation of historical data for improving accuracy in network regression. Stat Anal Data Min: ASA Data Sci J 10(1):40–53
31. Gao X, An H, Fang W, Huang X, Li H, Zhong W, Ding Y (2014) Transmission of linear regression patterns between time series: from relationship in time series to complex networks. Phys Rev E 90(1):012818
32. Holme P, Saramäki J (2012) Temporal networks. Phys Rep 519(3):97–125
33. Colliri T, Zhao L (2019) Analyzing the bills-voting dynamics and predicting corruption-convictions among Brazilian congressmen through temporal networks. Sci Rep 9(1):1–11
34. Csardi G, Nepusz T et al (2006) The igraph software package for complex network research. Int J, Complex Syst 1695(5):1–9
35. Thompson WH, Brantefors P, Fransson P (2017) From static to temporal network theory: applications to functional brain connectivity. Netw Neurosci 1(2):69–99
36. Câmara (2019) Dados Abertos. [Accessed on December 1, 2019]
37. Federal ST (2019) Processos. https://portal.stf.jus.br/. [Accessed on December 1, 2019]
38. Hulovatyy Y, Chen H, Milenković T (2015) Exploring the structure and function of temporal networks with dynamic graphlets. Bioinformatics 31(12):i171–i180
39. Liben-Nowell D, Kleinberg J (2007) The link-prediction problem for social networks. J Am Soc Inform Sci Technol 58(7):1019–1031
40. Guns R (2014) Link prediction. In: Measuring scholarly impact. Springer, Berlin, pp 35–55
41. Salton G, McGill MJ (1986) Introduction to modern information retrieval. McGraw-Hill Inc, New York
42. Spertus E, Sahami M, Buyukkokten O (2005) Evaluating similarity measures: a large-scale study in the Orkut social network. In: Proceedings of the eleventh ACM SIGKDD international conference on knowledge discovery in data mining, pp 678–684
43. Esquivel AV, Rosvall M (2011) Compression of flow can reveal overlapping-module organization in networks. Phys Rev X 1(2):021025
44. Egghe L, Leydesdorff L (2009) The relation between Pearson's correlation coefficient r and Salton's cosine measure. J Am Soc Inform Sci Technol 60(5):1027–1036
45. Ahlgren P, Jarneving B, Rousseau R (2003) Requirements for a cocitation similarity measure, with special reference to Pearson's correlation coefficient. J Am Soc Inform Sci Technol 54(6):550–560

46. Katz L (1953) A new status index derived from sociometric analysis. Psychometrika 18(1):39–43
47. Page L, Brin S, Motwani R, Winograd T (1999) The pagerank citation ranking: bringing order to the web. Technical report, Stanford InfoLab
48. Newman ME (2004) Fast algorithm for detecting community structure in networks. Phys Rev E 69(6):066133

Chapter 5
Networked Corruption Risks in European Defense Procurement

Ágnes Czibik, Mihály Fazekas, Alfredo Hernandez Sanchez, and Johannes Wachs

Abstract In this chapter we study corruption risks in EU defense procurement. Defense procurement has long been thought to present significant potential for corruption and state capture. Using a large dataset of contracts covering nearly ten years and applying an objective corruption risk indicator, we find strong empirical support for this hypothesis. In nearly all countries our corruption risk indicator is higher for military contracts than for contracts in general. By mapping national markets as complex networks, we find that risks are significantly clustered, suggesting potential islands of state capture. The centralization of corruption risk varies from country to country: in some corruption risk is significantly higher in the periphery, while in others it is significantly higher in the center of the market. We argue that network maps of procurement markets are an effective tool to highlight hotspots of corruption risk, especially in the overall high risk context of defense contracting.

5.1 Introduction

Public procurement is one of the government activities most vulnerable to corruption [23, 35]. Risks are even higher in the field of defence due to the large amounts of money involved, the complex and high-value contracts, high market concentration, and the fact that governments themselves are the enforcers of secrecy [26]. The

Á. Czibik
Government Transparency Institute, Budapest, Hungary

M. Fazekas
Central European University, Vienna & Government Transparency Institute, Budapest, Hungary

A. H. Sanchez
Institut Barcelona d'Estudis Internacionals, Barcelona and Government Transparency Institute, Budapest, Hungary

J. Wachs (✉)
Vienna University of Economics and Business & Complexity Science Hub Vienna, Vienna, Austria
e-mail: johannes.wachs@wu.ac.at

© The Author(s), under exclusive license to Springer Nature Switzerland AG 2021
O. M. Granados and J. R. Nicolás-Carlock (eds.), *Corruption Networks*,
Understanding Complex Systems, https://doi.org/10.1007/978-3-030-81484-7_5

defence procurement market has certain aspects which distinguish it from general public procurement, both in terms of market structure and regulation, which may limit efficiency and fair competition. While efficiency and quality of defence spending are of great importance for the public good via their impact on national security, citizens have limited options for monitoring and holding the government accountable in this field due to confidentiality, and a relative scarcity of publicly available information.

This chapter aims to gauge the extent and types of state capture in defence procurement across the EU. We go beyond measuring corruption risks by assessing the phenomenon of state capture drawing on recent complex network based approaches [8, 12, 34]. According to this conceptual and analytical framework, state capture is not just widespread corruption, but a tight clustering of corrupt actors and ties among them, typically centred around certain public organisations, government functions, or supply markets. Going beyond blanket averages of risks in countries or markets, measuring the distribution of corruption risk in procurement markets mapped as complex networks reveals a variety of ways in which corruption is organized.

This perspective on corruption risk has high relevance for anti-corruption policy, as captured clusters are expected to behave differently, thus demanding different solutions. Addressing state capture is especially relevant in defence procurement as the low number of contracting authorities and suppliers, the complex technology, typically large contract values and high degrees of secrecy in national security decisions create an environment of interdependence among insiders, and limit the capacity of outsiders to effectively monitor wrongdoing. Whether a high corruption risk cluster is central or peripheral in a country's military procurement market clearly has implications for underlying mechanisms and potential solutions.

To explore state capture in defence procurement we first apply a robust measure of corruption risk in public procurement transactions to a curated dataset of EU defence contracts. We report country-level corruption risk averages and compare them with non-military procurement outcomes, finding that military contracts tend to have higher corruption risk. To analyze the distribution of risks and assess potential clusters of corrupt capture, we construct a contracting network of organisations. We demonstrate that corruption risks are far from uniformly distributed in a majority of the markets we study. Researchers studying British and French military procurement have used our identification of key organizations that are central in their markets and have high corruption risk as a starting point for in-depth investigations into state capture [27, 28].

The rest of the chapter is organised as follows. We first provide an overview of the legal and economic factors which differentiate defence procurement from general procurement, including national security concerns, market structure, and the nuances of relevant EU legislation. We also review the findings of the literature addressing defence procurement in terms of market structure, corruption risks, and state capture.

We then describe the data sources we used to carry out the quantitative analysis, including the Corruption Risk Index we adapted from previous work [9], comparing military procurement risk with general procurement risk across the countries in our sample. We then proceed with a network analysis of these markets—highlight the highly non-random distribution of risk in most countries. We also consider the extent to which corruption risk appears in the center or periphery of various country markets. The last section summarises findings and formulates recommendations for policymakers and future research.

5.2 Defence Procurement in the EU

Although there is no single clear definition of defence procurement which is widely accepted by experts of the field, there is certainly a distinction between the products belonging to the very core of national security functions of the State—such as ammunition, submarines and vehicles for transporting troops—and the whole range of products acquired by authorities operating in the field of defence, which also includes goods and services necessary to fulfill administrative functions, such as office furniture and basic IT services. These two categories can be referred to as the narrow and the wide definitions of defence procurement [25]. The former covers goods and services which were manufactured or intended to be used for purely military purposes, especially armaments. Dual-use products and technologies can also be included if they were acquired for military use. The latter encompasses the totality of goods and services procured by entities related to national security.

The narrow and wide definitions of defence procurement draw attention to the fact that some goods and services in the field of defence are more affected by national security considerations than others. In this sense, the procurement of more sensitive goods requires a regulatory regime which acknowledges the defence-specific characteristics of this sector and finds the balance between openness and transparency of the procurement process on the one hand, and protection of the core security concerns on the other hand [25]. In contrast, the acquisition of non-sensitive defence-related supplies is quite similar to 'general' public procurement, so lack of transparency and restrictive procedures cannot be justified necessarily. This report focuses on sensitive goods and services in the field of defence, that is, the *narrow* definition of defence procurement. This means in practice that it is not the buyer but the product that determines whether we consider a tender as defense-related or not. For instance, we do not consider all purchases of ministries of defense as defense—related.

5.2.1 Defence Procurement Market Size

The 27 member states of the EU plus the UK spent 205 billion euros on defence in 2017 according to Eurostat, which is 1.7% of the GDP of these countries on average. However, this value covers several different types of expenses, such as salaries, foreign military aid, etc. so it cannot be used directly as an estimation of the total value of defence-related public procurement in Europe. The European Commission provides a method for the estimation of defence procurement in its working document *Evaluation of Directive 2009/81/EC on public procurement in the fields of defence*

Table 5.1 EU+UK and EU+UK+EEA annual defence procurement spending, rounded to the nearest billion Euro, sourced from Eurostat

	2007	2008	2009	2010	2011	2012	2013	2014	2015	2016	2017
EU28	79	81	79	79	78	80	80	81	90	91	94
EU28/EEA	83	86	84	84	83	85	85	86	95	97	100

and security which is based on 2010–2014 Eurostat data, where the total general government expenditure on military defence is further disaggregated into specific national accounts components. The maximum total value of military procurement can be estimated as the sum of 'Intermediate consumption' and 'Gross fixed capital formation'. The time series can be extended for the period 2007–2017 using the newest Eurostat data, see Table 5.1.

5.2.2   Defence Market: Buyers and Suppliers

Although in most countries the primary buyers of defence goods and services are ministries of defence, other types of entities also appear in this market, such as law enforcement and detention systems. While defence ministries are responsible for handling territorial threats and military crises, other institutions can be responsible for a wide range of tasks (e.g. combating terrorism or providing airport security). The number of potential buyers varies greatly among the subcategories of defence goods: only ministries of defence buy warships, but there are more potential buyers for firearms.

The suppliers of defence procurement are not clearly distinguishable from companies manufacturing "civilian" goods. Many companies, which produce goods for military use, have also other fields of activities without military character, and dual-use technologies are especially hard to be classified. In any case, two distinctions should definitely be considered when analysing defence procurement. First, there is an important difference between prime contractors—or system integrators—on the one hand, which are large companies capable of delivering complex security solutions (usually they are the ones signing contracts with buyers), and smaller companies on the other hand, which are usually subcontractors of the prime contractors. Naturally, these roles are not fixed, a middle-size firm may be the contracting partner of a buying entity in one transaction, and subcontractor in another; however, this flexibility strongly depends on the type of goods and the contract size. Second, the sensitivity of goods is an important factor. The market of core defence goods—or armaments—has certain characteristics which differentiates it from "civilian" markets, while this is less relevant in case of non-sensitive goods. To sum up, differentiating factors apply mostly for prime contractors operating in core defence markets, and these impacts fade gradually as we go deeper into the supply chain, entering the market of non-sensitive defence-related goods and services.

The total turnover of the defence industry sector was 97.3 billion euros in 2014, and 500 000 people were directly employed in this sector. However, defence capabilities are not evenly distributed among member states. EU countries can be classified into four broad groups based on their prime contractors and the size of their defence industry sector in general [33]. France and the UK are in the first group on account of their extensive defence industries, their status as nuclear powers, and their permanent seats in the UN Security Council. The second group contains countries with significant capacities: Germany, Italy, Sweden, and Spain, while the third group covers countries with limited capacities: Belgium, Finland, the Netherlands, Poland, Czech Republic, Romania, and Denmark. All other countries are in the fourth group with very limited or no defence capacities at all. It worth mentioning that even in countries with the largest defence industry, capacities are not enough to provide full range of equipment which results in a pressure for cooperation and mergers both at national and European level. This phenomenon has also consequences for competition: in case of expensive high-end technology such as aerospace technologies, competition is bound to be very limited, while competition can emerge in other sectors such as ships and vehicles.

Extensive supply chains are also an important characteristic of the defence industry, especially for complex contracts. The distribution of subcontractors is more even among EU countries than the distribution of prime contractors. Subcontracting is an opportunity for small and medium enterprises (SMEs) to participate in the defence industry.

National markets of certain goods and services are often characterised by monopsony, i.e. only one buyer on the market, and monopoly or oligopoly, i.e. only one or very few suppliers on the market, at the same time. The low number of actors, accompanied by protectionism, makes the relationship between governments and national champions often interdependent. This applies even more so to countries where the state has ownership in the biggest and strategically most important defence companies, e.g. in France, Portugal, Poland and Germany. Consequently, decisions regarding defence procurement depend not only on value for money and budget considerations, but industrial policy, employment, control over know-hows, and national security reasons, or any combination of these. This often leads to a setting in which the national champion has certain benefits that potentially distort competition, e.g. it is subject to tax exemptions, or contracts are awarded to it even if there would be other options.

At the European level, the defence market is characterised by fragmentation and duplication, which results in inefficiencies thanks to the lack of economies of scale. Inefficiency could mean not only higher prices but lower quality and longer completion time too, which could raise concerns regarding national security in the long term. In this sense, opening up the EU internal market for defence products is of high importance, which is addressed by a range of interventions, including Directive 2009/81/EC on defence and sensitive security procurement, however, there is still room for improvement.

5.3   Related Work

Competition, transparency and corruption risks are studied by academia as well as international think tanks and NGOs such as Transparency International Defence and Security Programme, the Stockholm International Peace Research Institute (SIPRI), and the Geneva Centre for the Democratic Control of Armed Forces (DCAF). These studies often use surveys and case studies from all around the world to illustrate problematic areas in military procurement and to recommend tools to tackle them. Survey data typically means that cross-country research makes use of corruption perceptions indices. Case studies are based on in-depth systematic data collection (both qualitative and quantitative) of selected events, organisations or countries. While the lack of broad scope mean that it is difficult to extrapolate to other settings, these research projects are still helpful in identifying key problems and vulnerabilities in defence procurement. Beyond exploring problems, advocacy organizations and think tanks usually draw up recommendations, that is, steps towards a solution: lowering corruption risks, more transparency, and better value for money.

In this section we survey related work on corruption risks in defence procurement. First we review work on how defence procurement has unique corruption risks, distinguishing it from other public sector activities. Next we describe the overall market structure of the defence sector, again suggesting ways in which its organization differs from that of other sectors. Finally, we review relevant literature on the measurement of corruption risk in procurement.

5.3.1   Distinguished Corruption Risks in Defence Procurement

Gupta et al. use aggregated budget data and corruption perception indicators to test the relationship between corruption and high levels of military spending in 120 countries in the period of 1985–1998 [16]. Their results indicate that corruption—measured by Transparency International's Corruption Perception Index and International Country Risk Guide Index—is indeed associated with higher military spending, measured by its share in both GDP and total government spending. This result supports the statement that military spending is associated with higher level of corruption risks compared to procurement in general, but it leaves open the question how corruption is done and what can be done to mitigate the risk.

According to Feinstein, Holden and Pace, the following built-in features of the arms trade make this field prone to corruption: (a) the secrecy related to national security and commercial confidentiality, (b) the close personal relationships between buyers, suppliers and their brokers, (c) the complexity, fragmentation, and often opacity of global production, transportation and financial networks, (d) the technical specificity of products, (e) procurement pressures, and (f) the high financial rewards coupled with a lack of consequences of wrongdoings [11]. Most of these factors appear also on the list of inherent risks and factors facilitating state capture in general [24], namely, technical complexity, opacity of decision making, stable policy networks with repeated interactions over time. This implies that besides one-off instances of corruption, state capture risks also have to be considered in the field of defence procurement.

Feinstein, Holden and Pace also describe the most frequently used methods to acquire undue influence in the arms trade, which are the following: (a) bribery (often through a third party which provides a legal remove between the supplier and the corrupting act), (b) failure to declare a conflict of interest, (c) the promise of post-employment, or revolving door, which blurs the line between the state and the defence industry and (d) the offer of preferential business access, which is often related to offsets, e.g. public officials are offered cheap or free shares in companies that have been founded in furtherance of an offset programme [11]. Most of these means (except for bribery) assume a stable, long-lasting relationship in the background, rather than a one-off transaction, which point at likely state capture in this field.

A comprehensive report of Transparency International's defence and Security Programme [17] explores the extent and the reasons behind non-competitive defence contracts in order to formulate recommendations for various actors in this field. They attempted to collect qualitative and quantitative defence procurement data from 45 defence ministries with special attention to non-competitive procedures, which they identified as a corruption risk in itself, but they only succeeded in seven countries, which in itself is a telling example of data challenges in this area. The countries participating in the research had single sourcing percentages between 9% (Bulgaria) and 55% (United Kingdom) in defence procurement, with even higher rates if we narrow down the analysis to armaments only. The following barriers to open competition were identified: (1) the protection of the national defence industry by over-using Article 346 of TFEU, (2) restrictive requirements in the request for tenders, (3) excessive use of classification, even in case of non-sensitive defence related information, (4) limited license rights, which often lead to a situation where repair and maintenance of an equipment can be done only by one contractor, i.e. the original supplier, (5) lack of unification of standards and interoperability of equipment.

Another report of TI UK analyses the corruption risks associated with defence offsets through three case studies [20]. Defence offsets are arrangements between the purchasing government and a supplier from another country, where the latter is obliged to invest a certain share of the contract in the importing country either through defence-related projects (e.g. by subcontracting), or through activities not related to defence such as purchases of other goods and services. The percentage of the offsets contract is often very high, even above 100%, and they are highly susceptible to corruption due to their complexity and a reduced level of scrutiny compared to the main arms deal. The study identifies three main categories of corruption risks from offsets: (1) influencing the need for a particular defence acquisition, (2) influencing the decision for the main contract, (3) allowing favours to be repaid to corrupt government officials via the offset contracts.

5.3.2   Defence Procurement Market Structure

The analysis of market structure complements the assessment of potential state capture in the field of defence procurement. The markets in which buyers and suppliers are embedded can both influence behavior and reflect existing arrangements. Factors such as market concentration and buyer centralization are often studied theoretically, for instance via principal-agent models [21] or by models of competition [14] or by contrasting auction formats [5]. Recognizing that defense markets often mix and match sub-market types (monopsony, oligopoly, etc.), we focus on the empirical structure of these markets, observing them as they are and relating them to risk indicator outcomes. Nevertheless, it is important to give an overview of the general patterns of structure in these markets.

The relationships of companies in the defence industry is often described as a hierarchy of 'tiers'. Prime contractors (or 'primes') are on the top of this pyramid. They are specialised in defence production and sell complex products, such as weapon systems to the end users, i.e. mainly government agencies and ministries of defence. Below that is the first tier containing system providers, who are the producers of complete subsystems or major components. They are the final step before the product reaches the prime contractor, who may complete the product or simply organise the shipment, marketing, etc. Below the first tier, there are second tier and third tier companies, often producing dual-use components for military purposes after being integrated into larger systems. They are not always listed as defence producers because they usually produce non-defence goods too. Most academic studies exploring European defence market structures focus on prime contractors, and the consolidation process at European-level. Very little evidence is available on first-tier, second tier (and lower tier) companies and the market processes at the national level.

Carril and Duggan analyse the impact of increasing concentration of the 1990's US defence market on procurement outcome variables [2]. Using micro-level data (US's Department of defence contract awards), they find that market concentration made the procurement process less competitive, which was evidenced by the increasing share of contracts awarded without competition, or via single-bid solicitations. Contracts tended to shift from fixed-price towards cost-plus contracts. However, they found no evidence that consolidation led to a significant increase in acquisition costs of large weapon systems, neither to increased spending at the product market level. The government's buyer power constrained firms from exercising any additional market power gained from consolidation.

The structure of the defence market is analysed from a political-business perspective by Neil and Taylor who describe different paths of restructuring after the Cold War in the United States and Europe, focusing on prime contractors [22]. They show that while the major approach of consolidation in the US was merger and acquisition, in Europe, more cautious approach was applied, which consisted of a wide range of tools for consolidation such as strategic alliances, minority shareholdings, and joint ventures. The study states that whilst the core drivers of consolidation were similar in the US and Europe, the more complex relationship-system of European defence companies, which emerged due to the many national champions involved, may be an advantage in the global market, where flexibility and the ability to deal with cultural and political differences have great significance.

To sum up, there is evidence in the literature that defence procurement is especially prone to low level of competition, lack of transparency and corruption risks compared to 'general' procurement. Recent case studies of the British and French defence procurement markets confirm that these risks manifest in complex ways [27, 28]. The reasons include, on the one hand, the extensive use the notion of national security which limits the usability of usual monitoring mechanisms; on the other hand, the size, complexity and technical specificity of major arms programmes making hiding corruption relatively easy. An empirical study of the Spanish defense industry highlights great heterogeneity in performance and efficiency within a national market [6]. The level of competition and the power relations among buyers and suppliers strongly depend on the specific product and market: corruption risk likely varies across subsectors as well.

5.3.3 Measuring Corruption Risk

Having established that the defence industry, in particular its procurement arm, merits examination, we turn to the topic of measuring corruption risks. An emerging field of research quantifies corruption risks in public procurement using contract-level indicators that track the extent to which a contract's award deviated from a norm of free and impartial competition [9].

For example, a contract awarded directly to a firm without competition clearly present a corruption risk. Competitors are often excluded in subtle ways, for instance by onerous and specific requirements for past experience or by the imposition of an impossibly short deadline to submit tenders. In these cases, the requirements might be tailored to the favored firm and the firm can be tipped off ahead of time about the call for tenders. Pooled together into a composite measure, these indicators provide a fine-grained, data-driven proxy for corruption risk in procurement contracts. Aggregated to regional and national levels, these indicators have a strong correlation with generally accepted measures of corruption prevalence [7]. They have been used to evaluate, among other things, the effectiveness of meritocratic promotion in improving quality of government [3].

Another great advantage of these micro-level measures of corruption risk is that they offer the opportunity to study the distribution of corruption risks within a country, market, or region. By mapping procurement markets as bipartite networks of contracting among buyers and suppliers, one can go beyond averages and study the complex organization of corruption. In general, it is known that corruption risks predict missing contracting edges, suggesting that corruption is about exclusion of non-favored competitors [10]. Corruption risks are also reflected in market structure across politically meaningful elections: network neighborhoods of high corruption risk actors rewire significantly across change in government [10]. While corruption risk indicators alone cannot prove corruption has taken place, they offer an alternative perspective on this important issue by searching for traces of organized bad behavior in broad data.

The network perspective has also given new insight into the phenomenon of capture of specific parts of a state by corrupt actors [12]. This manifests as clusters of highly corrupt actors in procurement market networks [34]. When such clusters exist in the center of a country's procurement network, this suggests that the situation is especially dire [8]. Altogether, this line of research reflects a growing recognition that corruption and economic crime in general is organized among many individuals and actors in a complex way [1, 18, 19, 30, 31].

5.4   Data

In this section we outline the data sources we used for our analysis and the major steps we took to prepare the data for analysis.

5.4.1   Tenders Electronic Daily—TED

We collected contracts from a centralized database known as Tenders Electronic Daily (TED), the official EU portal for contract notices and awards. On the site, contracting authorities publish their calls for tenders and contract award notices above certain value thresholds, which differs for goods, services and works. Notices on TED contain the most important pieces of information on the tendering process such as: the title and description of the tender, publication date and bidding deadline, estimated and final value, information on the tendering procedure and the identity of the buyer and the winner. Before we could use this dataset for analysis, entity deduplication was necessary. Available public contracting data does not typically assign unique identifiers to entities involved in the contracting process. In other words, buyers and suppliers of goods and services are identified by plain text names and not tax numbers. For example, a contract awarded by the British Ministry of defence to BAE Systems may list "MoD" as the buyer, and "BAE Systems, Ltd." as the supplier. Another contract between the same two entities may list "Ministry of defence" and "BAE Systems". In order to properly analyse these markets, it is important to identify and merge the aliases of both buyers and suppliers as accurately as possible.

Following deduplication, we considered all awarded contracts from 2006 to 2016, and filtered the data for contracts pertaining to defence-related activities. There are two ways in which we label a contract as military-related: a) one of the Common Procurement Vocabulary (CPV) product codes listed in the tender documentation comes from a list of curated codes deemed military related (see Appendix B on CPV codes), or b) the contract falls under the purview of the EU Directive 2009/81/EC on defence

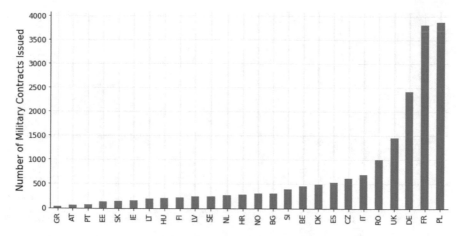

Fig. 5.1 Number of military-related procurement contracts per country

and sensitive security procurement. The resulting dataset contains 18,608 contracts. We plot the count of military contracts in our database in Fig. 5.1. Unsurprisingly, we generally have more contracts from larger countries. In a parallel effort to collect data on procurement from media reports, we found a significant amount of procurement contracts that were missing from the data, suggesting a major transparency failure, see: [4].

5.5 Measuring Corruption Risks

To quantify the corruption risks at the contract level, we adapt two objective corruption risk indicators from the academic literature [9]. Such indicators count "red flags" in how a contract was awarded, capturing competition or transparency-limiting tricks that have been used to steer contracts to preferred winners. The first contract level indicator is single bidding: did the contract attract only a single offer from the private sector? This indicator considers only the outcome: whether there was competition for the contract.

The second indicator we consider is a composite index of red flags. In addition to the single bidding rate, we consider:

- Procedure type: was the contract not awarded by an open competition (i.e. by direct negotiation or by an invitation-only procedure)?
- Length of advertisement period: was the time to submit bids notably short?
- Evaluation criteria: to what extent were the bid evaluation criteria subjective (i.e. referring to unmeasurable notions of quality rather than objective criteria such as price, length of warranty, etc.)

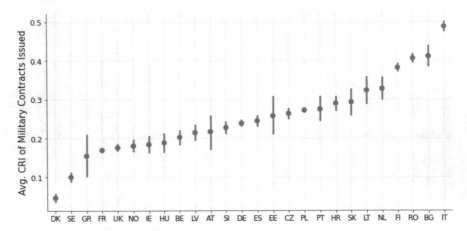

Fig. 5.2 Average Corruption Risk Index (CRI) scores on military contracts by country, from 2008–2016. The CRI tracks the presence of six red flags in the contract award process. Error bars represent 95% bootstrapped confidence intervals

- Call for tender publication: was the call for bids published in the official national or European procurement journal?
- Length of decision period: was the duration of the decision period either very short (indicating a premediated decision) or very long (indicating possible legal challenges)?

We count the number of red flags for each contract (and divide by 6) to arrive at its Corruption Risk Index (CRI). For instance, a contract awarded to a single bidder with a very short time to submit bids would have a score of 2/6. The CRI has been amply used in the literature on corruption in public tenders. Fazekas and Kocsis find that contract CRI scores tend to be higher for contracts awarded to winners registered in tax havens (2009–2014) [7]. Similarly, they find that single-bidder and high CRI contracts are associated with higher prices. This indicator directly captures corruption as unwarranted barriers to entry to privilege well-connected contractors in detriment of potential competitors. We plot the average CRI scores with bootstrapped 95% confidence intervals for defence procurements in Fig. 5.2.

For both indicators, we observe significant heterogeneity in corruption risks across countries. In Denmark, less than 1 in 10 military-related procurement contracts are awarded to a single bidder while in Italy, every second military contract is awarded in this way. While the overall picture confirms typical rankings of corruption and quality of government in EU countries [34], there are important exceptions. Finland usually ranks in good governance rankings—here we observe that Finnish military procurement contracts are often awarded with many red flags. Finland's unique geopolitical history as independent state between NATO and the USSR suggests that many military suppliers are Finnish. The Finnish state, like the Latvian one, may prioritize the onshore presence of critical suppliers over competitive market outcomes. Greece on the other hand has a relatively good scoring defense procurement market.

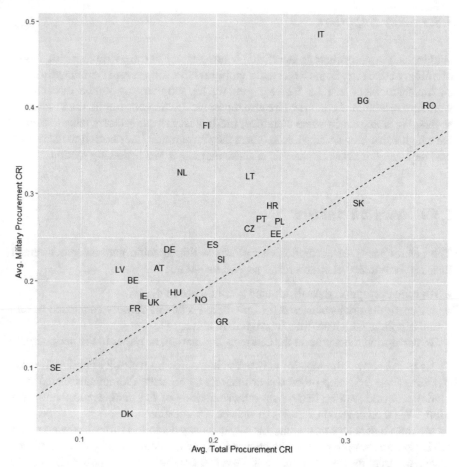

Fig. 5.3 Comparing average CRI scores for all contracts versus only military contracts, by country

To better understand how military procurement differs from procurement in general in, we plot the average CRI score of each country's entire procurement market (including traditional products such as road repair, medicine, school lunches) against their average military procurement CRI in Fig. 5.3. This provides us with a baseline for comparisons, highlighting again the Finnish, Greek, and Latvian cases as interesting outliers against the trend.

We draw two conclusions from this plot. The first is that in most countries military procurement contracts have higher corruption-risk scores than other contracts, most countries are above the 45-degree line. Second, there are significant outliers, indicating that military procurement carries significantly more (or less) corruption risk in certain countries. In Italy, Bulgaria, Finland and the Netherlands, military procurement has significantly higher corruption risks than procurement in general. The opposite is true in Denmark: there military procurement contracts have less corruption risk than other kinds of procurement, on average.

5.5.1 Key Suppliers

Within each country there is significant heterogeneity in the corruption-risk scores of military contracts. Some buyer and supplier relationships seem significantly more corrupt than others. In Table 5.2 we present the top suppliers, by number of contracts won, for a selection of countries. We also report their average corruption risk indicator scores. We note that in some countries, the largest private sector winners seem to have high corruption risks, while in others they have rather low corruption risks. This will motivate our network analysis of these markets in the following section.

5.5.2 Network Analysis

The heterogeneity of corruption risk scores within specific national procurement markets for defence contracts raises several questions:

- How are corruption risks distributed within these markets?
- Are corruption risks clustered (i.e. are there groups of densely connected buyers and suppliers which are more corrupt than average)?
- Do corruption risks arise in the centre of the market, or rather in the periphery?

As noted above, the tools of network science can be fruitfully applied to these questions. We first map procurement markets as bipartite (sometimes called two-mode) networks, noting that visual representations of the markets are themselves useful. We then develop measures to answer our questions.

We map defence procurement markets as networks in the following way: nodes are buyers and suppliers of public contracts. They are connected by an edge if they have a contracting relationship, i.e. if buyer A contracts with supplier Z, they are connected in the network. Mathematically, we describe the markets as bipartite graphs G, consisting of two kinds of nodes B, the buyers, and S, the suppliers of military contracts. An buyer $b \in B$ is connected by an edge $e_{b,s}$ with a supplier $s \in S$ if they have a contracting relationship.

In the visualizations below, gold nodes are buyers and black nodes are supplier. We colour the edges red if the average CRI of the contracts between the two nodes is at least one standard deviation above the market average. The nodes are placed using a physics-inspired algorithm: nodes are treated as charged particles which repel each other, while edges act as springs, pulling connected nodes closer to each other [13]. We visualize three national markets: Italy, the UK, and Germany in Fig. 5.4. These are among the larger markets in our dataset and cover a range of corruption risk outcomes. The regularities we observe in their network structure suggests how we might compare all of the countries in our dataset using network-measures.

We can draw a few qualitative conclusions from the networks of Italy, UK, and Germany. The first is that corruption risks seem to be clustered: red edges seem more prevalent in certain parts of the network than others. The second is that corruption

Table 5.2 Top winners of defence contracts in Italy, the UK, France, and Germany by number of contracts. Note: When data for CRI or Single Bidder is unavailable (NAs), we impute the country average. The assumption being that lack of information of on a given tender implies that its corruption risk is at least at the level of the country's average

Winner name	No. of contracts	Avg. CRI*	Single bidding rate*
Italy			
Agustawestland Spa.	30	0.56	0.68
Selex Es Spa.	20	0.53	0.52
Oto Melara Spa.	17	0.49	0.49
Piaggio Aero Industries Spa.	13	0.56	0.77
Alfredo Grassi Spa.	12	0.39	0.17
UK			
Mott McDonald Limited	23	0.05	0
Ch2M Hill United Kingdom	20	0.05	0
Lion Apparel System Limited	20	0.03	0.009
Hunter Apparel Solutions Ltd.	18	0.02	0
Parsons Brinckerhoff Ltd.	17	0.05	0
France			
Lognavcm	78	0.08	0.01
Balsan	54	0.16	0.25
Mainco	50	0.09	0.003
Gk Professional	49	0.14	0.16
P Poinsot	40	0.11	0.08
Germany			
Kraussmaffei Wegmann Gmbh. Co. Kg.	83	0.30	0.67
Rheinmetall Landsysteme Gmbh.	76	0.2	0.3
Ffg Flensburger Fahrzeugbau Gesellschaft Mbh.	63	0.19	0.1
Ruag Ammotec Gmbh.	59	0.32	0.51
Scharrer Konfektions Gmbh. Co. Kg.	44	0.17	0.07

Fig. 5.4 The military procurement networks of Italy, the UK, and Germany. Gold nodes are buyers and black nodes are suppliers. Edges indicate contracting relationships, with red edges highlighting relationships in which the average Corruption Risk Index (CRI) score is at least one standard deviation above the average in that country. We observe that high corruption risk edges appear to cluster together

risks appear more common near the centre of the network. Finally, in all three countries there are different types of buyers: some are hubs issuing contracts to many suppliers, while others issue contracts to only a few suppliers. To make these notions more precise, we can use methods to quantify the clustering and centralization of corruption risks in procurement markets mapped as networks.

To calculate the clustering of corruption risk we calculate the average correlation of an edge's CRI with that of its neighbours. In other words, we quantify the extent to which knowing one edge's CRI allows us to predict the CRI of neighbouring edges. If the correlation is high, it means that neighbours of high CRI edges are more likely to have high CRI, and vice versa. Mathematically, given an edge $e_{b,s}$ representing a contracting relationship between buyer b and supplier s, we first calculate the average CRI of the contracts between b and s. We then consider the adjacent edges, i.e. those edges who have either b as buyer or s as supplier and average the CRI across those edges. We calculate the Pearson ρ correlation between these two CRI scores.

We normalize the correlation using a permutation test, to enable comparisons between countries. In particular, we recalculated the edge-CRI correlation after shuffling the CRI outcomes across contracts. Repeating this one hundred times, we calculated a Z-score for each market, subtracting the observed correlation from the average correlation under randomization, then dividing by the standard deviation of the correlation under randomization.

In Fig. 5.5 we see that in most defence procurement markets, corruption risks are significantly clustered. This is especially true in the larger markets. This confirms our intuition from the network diagrams: if you find one red edge (a high corruption risk relationship), it is likely that edges around that buyer node will also be red. This is in line with our expectations that corruption risks are not randomly distributed across buyer-supplier relationships, but rather clustered around key institutions— see Fazekas and Toth [8].

To quantify the idea that corruption risks seem more prevalent at the centre of the market, we calculate the so-called closeness centrality of each buyer and relate

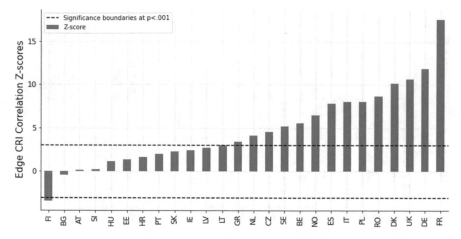

Fig. 5.5 The edge-clustering of corruption risk in different countries. We calculate correlation of corruption risk on neighboring edges in each country's procurement market, and compare it against a shuffled null model

this with the average CRI of the contracts it issues. Closeness centrality is inversely proportional to a node's distance to all other nodes in the network. If one node is close to many other nodes, it is in some sense central in the network, while if it is very far from other nodes, it is in the periphery. Mathematically, we calculate the closeness centrality C of a node x as:

$$C(x) = \frac{N}{\sum_{y \in G} d(y, x)}$$

where d denotes the network distance between two nodes. We focus our attention on the buyers, the public institutions issuing defense contracts. We calculate the Pearson ρ correlation of their average CRI scores with their closeness centrality score, seeking to quantify whether buyers in the center of the network are issuing more or less risky contracts.

In Fig. 5.6 we plot these correlations. We find that in some countries such as the Netherlands, Finland, Slovenia and Germany, corruption risk is more prevalent in the centre of the market (indicated by a high correlation between buyer closeness and CRI). There are also countries where corruption risk is more prevalent in the periphery of the market such as Greece, Portugal and Estonia. This again highlights the non-uniform distribution of corruption risks in these markets.

In summary, network science methods enable us to map public procurement markets in an interesting way. They can also help us quantify intuitions about the distribution of corruption risk in a market. We find that in general, corruption risk is clustered, indicating systematic state capture rather than a random phenomenon. On the other hand, corruption risk is not always more common in either the center or

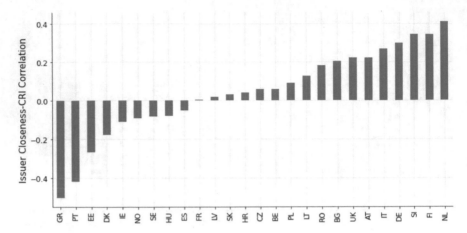

Fig. 5.6 The correlation between closeness centrality of buyers and their corruption risk scores by country. A high correlation suggests that corruption risk is higher in the center of the network

periphery of a market. In some countries corruption risk is more common among core institutions, while in others it thrives on the periphery.

Taken together, our framework generates a natural typology of corruption risks in defense markets. For instance: corruption risks in Finnish defense procurement contracting are relatively high, unclustered, and tend to appear more in the center of the network than its periphery. The Greek market has relatively low corruption risks, a moderate amount of clustering, and has higher risks in its periphery than among its central institutions. While a detailed comparison of different markets is beyond the scope of the chapter, these findings suggest how our framework can be applied to understand the distribution of risks.

5.6 Discussion

In this chapter we carried out a quantitative analysis of corruption and state capture risks in the field of defence procurement in Europe. First, we collected data using official and alternative sources to tackle the challenges typical for this sector, namely the relatively low level of transparency compared to most other procurement markets. We found that the use of alternative sources such as news articles is ambiguous: while the lack of exact details limits the usability of such additional data for research purposes; media often reports on the large value strategic purchases which are typically not published on official tendering websites. As a consequence, alternative sources cover a larger share of total defence procurement expenditure than notices published on the official platform in many countries. In this sense, they increase transparency significantly, and they raise public interest towards defence procurement, which creates a pressure to publish better, more comprehensive official datasets in the long run.

We analysed the large database of contracts collected from Tenders Electronic Daily from several perspectives. We began by identifying the typical corruption risk in defence contracting, finding great heterogeneity across EU countries. For instance, while roughly every other military contract awarded in Italy from 2006 to 2016 was awarded to a single bidder, only one in twenty contracts in Denmark were awarded in such a way. This reflects the situation in public procurement more generally, though it is in some sense surprising given that military procurement is high profile and perhaps more internationally relevant than procurement of local roads or health services.

Within-countries, we observed a significant positive correlation between corruption risk in the military procurement sector and corruption risks in procurement more generally. In other words, corruption risks in military procurement closely reflect corruption risk patterns at the national level. Overall, military procurement risks are higher than other procurement sectors in nearly all European countries. The largest corruption risk premiums in military procurement over risk in other kinds of procurement exist in Italy, the Netherlands, Finland, and Bulgaria.

A significant advantage of measuring corruption risk using contracting data is that it enables micro-level analyses of key actors. By listing the corruption risk scores of top winning firms in different countries, we observed that distribution of corruption risks within countries can be quite heterogeneous. For instance, while the overall corruption risk rate of contracts awarded in Germany was moderate, some frequent winners had single bidding rates of over 50%, while others had single bidding rates below 10%. In Italy on the other hand, nearly all of the top winners had single bidding rates above 50%. This suggests that corruption risks are not randomly distributed in different markets. The findings are used to identify certain high-risk networks of buyers and suppliers where detailed field research was carried out in order to explore them in more detail. The results of two such case studies using our findings indicate how network analysis can complement qualitative investigations [27, 28].

We took another look at the distribution of corruption risks across the contracting relationships between buyers and suppliers using network analysis. By visualising the markets as networks, we could demonstrate more clearly what we claimed before: that corruption risks are not random, but rather clustered in the relationships of distinguished buyers and suppliers. Such networks offer analysts and the authorities a bird's eye view of the distribution of corruption risks in the market and state capture by implication. It also offers a framework to quantify the nature of corruption in a given market, for instance if it is more often present in the centre of a market or in its periphery. We found examples of both kinds of markets, underscoring that corruption risks manifest themselves in different ways in different countries. We argue that a network map of markets provides a useful tool to understand these complex differences both at a glance and with a view to investigate them further. Even more broadly, networks are able to highlight important emergent properties of economic interactions embedded in social and political life [29, 32].

We highlight several avenues for future work, building on our findings and methods. The most obvious way to generalize our work is to expand the number of countries considered. Certainly our work was facilitated by the existence of a cross-country comparable data portal for procurement awards in the EU. While similar data sources exist for countries outside the EU, making them comparable (owing to varying regulations and reporting thresholds, for example) is a considerable task. That said, our framework could be applied to these interesting cases. As a share of GDP, military spending is significantly higher in Russia, China, and the US compared to the EU and UK. Emerging market countries are also major customers of defense firms. Tracking the behavior of multinational defense contractors in these markets is one possible direction. Here again networks can provide substantial value: analysis of financial transactions, ownership structures and board members can yield surprising insights [15].

To sum up, network methods are an effective monitoring tool, as well as a quantitative framework to understand the organization of corruption in procurement markets. As corruption and more generally state capture are phenomena which cannot be neatly characterized as either entirely micro or macro, network analysis is a useful lens through which they can be observed.

Acknowledgements We thank the participants of the workshop "Defence procurement in the EU and data" organised at the University of Nottingham, especially Aris Georgopoulos. Acknowledge support from Open Society Foundations under the grant "State capture and the defence procurement in the EU" (grant number: OR2017-34852).

References

1. Campedelli GM (2020) Where are we? using scopus to map the literature at the intersection between artificial intelligence and research on crime. J Comput Soc Sci, 1–28
2. Carril R, Duggan M (2020) The impact of industry consolidation on government procurement: Evidence from department of defense contracting. J Public Econ 184:104141
3. Charron N, Dahlström C, Fazekas M, Lapuente V (2017) Careers, connections, and corruption risks: Investigating the impact of bureaucratic meritocracy on public procurement processes. The Journal of Politics 79(1):89–104
4. Czibik Á, Fazekas M, Sanchez AH, Wachs J (2020) State capture and defence procurement in the EU. Technical report 2020:05, Government Transparency Institute
5. Dana JD Jr, Spier KE (1994) Designing a private industry: Government auctions with endogenous market structure. J Public Econ 53(1):127–147
6. Duch-Brown N, Fonfría A, Trujillo-Baute E (2014) Market structure and technical efficiency of spanish defense contractors. Defence and Peace Economics 25(1):23–38
7. Fazekas M, Kocsis G (2020) Uncovering high-level corruption: cross-national objective corruption risk indicators using public procurement data. British Journal of Political Science 50(1):155–164
8. Fazekas M, Tóth IJ (2016) From corruption to state capture: A new analytical framework with empirical applications from hungary. Polit Res Q 69(2):320–334
9. Fazekas, M., Tóth, I.J., King, L.P.: An objective corruption risk index using public procurement data. European Journal on Criminal Policy and Research 22(3), 369–397 (2016)
10. Fazekas M, Wachs J (2020) Corruption and the network structure of public contracting markets across government change. Politics and Governance 8(2):153–166
11. Feinstein A, Holden P, Pace B (2011) Corruption and the arms trade: sins of commission. SIPRI (2011)
12. Fierascu S (2019) Redefining state capture: the institutionalization of corruption networks in Hungary. Eikon Bucharest

13. Fruchterman TM, Reingold EM (1991) Graph drawing by force-directed placement. Softw: Pract Exper 21(11):1129–1164
14. García-Alonso, M.D., Levine, P.: Strategic procurement, openness and market structure. International Journal of Industrial Organization 26(5), 1180–1190 (2008)
15. Garcia-Bernardo J, Fichtner J, Takes FW, Heemskerk EM (2017) Uncovering offshore financial centers: Conduits and sinks in the global corporate ownership network. Sci Rep 7(1):1–10
16. Gupta, S., De Mello, L., Sharan, R.: Corruption and military spending. European Journal of Political Economy 17(4), 749–777 (2001)
17. Transparency International (2017) Single sourcing: a multi country analysis of non-competitive defence procurement
18. Kertész J, Wachs J (2020) Complexity science approach to economic crime. Nat Rev Phys, 1–2
19. Luna-Pla I, Nicolás-Carlock JR (2020) Corruption and complexity: a scientific framework for the analysis of corruption networks. Applied Network Science 5(1):1–18
20. Magahy B, da Cunha FV, Pyman M (2010) Defence offsets: addressing the risks of corruption and raising transparency. Transparancy International-UK
21. Mahoney CW (2017) Buyer beware: How market structure affects contracting and company performance in the private military industry. Secur Stud 26(1):30–59
22. Neal DJ, Taylor T (2001) Globalisation in the defence industry: an exploration of the paradigm for us and european defence firms and the implications for being global players. Defence and Peace Economics 12(4):337–338
23. OECD: Preventing corruption in public procurement (2016)
24. OECD: Preventing policy capture: integrity in public decision making. OECD public governance reviews (2017)
25. OECD-SIGMA: Defence procurement (2011)
26. Pyman M, Wilson R, Scott D (2009) The extent of single sourcing in defence procurement and its relevance as a corruption risk: A first look. Defence and Peace Economics 20(3):215–232
27. Renon E (2019) Does the defence industry capture the state in France? Technical report, 2019:03, Government Transparency Institute
28. Resimic M (2019) Public-private relationships in defence procurement in the EU: the case of the UK. Technical report 2019:04, Government Transparency Institute
29. Resimic M (2021) Network ties and the politics of renationalization: Embeddedness, political-business relations, and renationalization in post-milosevic Serbia. Comp Pol Stud 54(1):179–209
30. Ribeiro HV, Alves LG, Martins AF, Lenzi EK, Perc M (2018) The dynamical structure of political corruption networks. Journal of Complex Networks 6(6):989–1003
31. Smith CM, Papachristos AV (2016) Trust thy crooked neighbor: multiplexity in chicago organized crime networks. Am Sociol Rev 81(4):644–667
32. Stark D, Vedres B (2012) Political holes in the economy: The business network of partisan firms in hungary. Am Sociol Rev 77(5):700–722
33. Trybus M (2014) Buying defence and security in Europe. Cambridge University Press, Cambridge
34. Wachs J, Fazekas M, Kertész J (2020) Corruption risk in contracting markets: a network science perspective. Int J Data Sci Anal, 1–16
35. World Bank (2014) Fraud and corruption awareness handbook: a handbook for civil servants involved in public procurement

Chapter 6
Identifying Tax Evasion in Mexico with Tools from Network Science and Machine Learning

Martin Zumaya, Rita Guerrero, Eduardo Islas, Omar Pineda, Carlos Gershenson, Gerardo Iñiguez, and Carlos Pineda

Abstract Mexico has kept electronic records of all taxable transactions since 2014. Anonymized data collected by the Mexican federal government comprises more than 80 million contributors (individuals and companies) and almost 7 billion monthly-

M. Zumaya
Programa Universitario de Estudios sobre Democracia, Justicia y Sociedad & Centro de Ciencias de la Complejidad, Universidad Nacional Autónoma de México, 01000 Ciudad de México, Mexico
e-mail: martin.zumaya@puedjs.unam.mx

R. Guerrero
Plantel Del Valle, Universidad Autónoma de la Ciudad de México, 03100 Ciudad de México, Mexico

E. Islas
Subsecretaría de Fiscalización y Combate a la Corrupción, Secretaróa de la Función Pública, 01020 Ciudad de México, Mexico

O. Pineda
Azure Core Security Services, Microsoft, Redmond, WA 98052, USA

Posgrado en Ciencia e Ingeniería de la Computación & Centro de Ciencias de la Complejidad, Universidad Nacional Autónoma de México, 01000 Ciudad de México, Mexico

C. Gershenson (✉)
Instituto de Investigaciones en Matemáticas Aplicadas y Sistemas & Centro de Ciencias de la Complejidad, Universidad Nacional Autónoma de México, 01000 Ciudad de México, Mexico
e-mail: cgg@unam.mx

Lakeside Labs GmbH, 9020 Klagenfurt am Wörthersee, Austria

G. Iñiguez
Department of Network and Data Science, Central European University, 1100 Vienna, Austria
e-mail: iniguezg@ceu.edu

Department of Computer Science, Aalto University School of Science, 00076 Aalto, Finland

Centro de Ciencias de la Complejidad, Universidad Nacional Autónoma de México, 01000 Ciudad de México, Mexico

C. Pineda
Instituto de Física, Universidad Nacional Autónoma de México, 01000 Ciudad de México, Mexico

89

aggregations of invoices among contributors between January 2015 and December 2018. This data includes a list of almost ten thousand contributors already identified as tax evaders, due to their activities fabricating invoices for non-existing products or services so that recipients can evade taxes. Harnessing this extensive dataset, we build monthly and yearly temporal networks where nodes are contributors and directed links are invoices produced in a given time slice. Exploring the properties of the network neighborhoods around tax evaders, we show that their interaction patterns differ from those of the majority of contributors. In particular, invoicing loops between tax evaders and their clients are over-represented. With this insight, we use two machine-learning methods to classify other contributors as suspects of tax evasion: deep neural networks and random forests. We train each method with a portion of the tax evader list and test it with the rest, obtaining more than 0.9 accuracy with both methods. By using the complete dataset of contributors, each method classifies more than 100 thousand suspects of tax evasion, with more than 40 thousand suspects classified by both methods. We further reduce the number of suspects by focusing on those with a short network distance from known tax evaders. We thus obtain a list of highly suspicious contributors sorted by the amount of evaded tax, valuable information for the authorities to further investigate illegal tax activity in Mexico. With our methods, we estimate previously undetected tax evasion in the order of $10 billion USD per year by about 10 thousand contributors.

6.1   Introduction

Tax has a crucial role in the economic growth and welfare of the general population. Paid taxes allow for government spending and public expenditures (in the short, medium, and long term) such as education, healthcare, housing, pensions, security, and infrastructure [1]. Tax collection in Mexico is determined by a series of laws (*Ley de Ingresos de la Federación*, in Spanish) specifying the eligibility of taxpayers, tax rates and types, as well as the periodicity of fees and means of payment.

Despite the fact that taxes allow the Mexican state to fund beneficial activities for society, some individuals, corporations, or trusts may decide to illegally evade taxes by misrepresenting their state of affairs to the Tax Administration Service in Mexico (*Servicio de Administración Tributaria*, or SAT). In this sense, tax evasion is the reduction of constitutional tax liability by dishonest reporting, such as understating financial gains or overstating deductions [2].

Since 2014, Mexico has kept electronic records of all taxable transactions by means of a digital receipt or invoice known as *Comprobante Fiscal Digital por Internet* (CFDI). Each of these mandatory receipts includes data on the product or service transferred between taxpayers, date of transaction, cost, and corresponding tax amount. CFDIs are XML documents with technical specifications updated annually by SAT, including a certification seal that can only be produced by authorized parties [3]. The CFDIs are an integral part of formal investigations by SAT in tax evasion, money laundering, and other tax-related illegal activities, because they can potentially uncover the networks of individuals and legal entities involved in them.

Among all forms of tax evasion, here we focus in situations where taxpayers issue CFDIs in the absence of actual economic activity to increase tax deductions. Such legal entities, known as 'enterprises billing simulated operations' ('*empresas que facturan operaciones simuladas*', or EFOS), are typically characterized by a lack of employees or infrastructure, as well as a fake or constantly changing address used to avoid detection. As a result of investigations already led by SAT, EFOS can be classified as *definitive* or *alleged* tax evaders [4]. In order to operate, EFOS require the participation of 'enterprises deducting simulated operations' ('*empresas que deducen operaciones simuladas*', or EDOS). Even if EDOS engage in illegal activity alongside EFOS, they tend to have demonstrable stability in their workforce, assets, and tax contributions. By receiving CFDIs associated with simulated operations, EDOS aim to reduce their tax rate to avoid payments to SAT and eventually obtain further fiscal benefits.

In order to decrease the risk of systematic tax evasion, the Mexican federal government has established fiscal law (known as '*Código Fiscal de la Federación*') describing the official procedure to identify taxpayers as EFOS: (a) First, SAT determines the lack of actual economic activity behind a set of CFDIs. (b) Then, the government notifies the relevant legal entities via its official publication ('*Diario Oficial de la Federación*'). (c) Alleged EFOS have 15 days to contest the claim. (d) Associated EDOS (that have received the suspicious CFDIs) can correct their standing with SAT by resubmitting appropriate tax forms. (e) Finally, if any tax revenue has been lost to illegal activity, SAT classifies relevant taxpayers as definitive EFOS and specifies the type of crime following fiscal law. Definitive EFOS are unable to emit further CFDIs.

Despite its effectiveness, the official procedure of detecting tax evasion is time and resource consuming, particularly in initial stages of the process where suspicious CFDIs and alleged EFOS need to be manually selected from millions of contributors and billions of transactions each year. In order to complement these efforts, automated computational and statistical techniques (with tools from network science and machine learning) can be used to characterize the network of issued/received CFDIs between definitive EFOS and other taxpayers. By analyzing the temporal properties of the network of all taxable transactions in Mexico from 2015 to 2018, here we show evidence of a group of highly suspicious contributors with behavior statistically similar to that of definitive EFOS, as well as estimates of previously undetected tax evasion. When combined with current practices at SAT, this information has the potential to increase the efficacy of the governmental response to illegal tax activity in Mexico.

6.2 Data

Taxpayers in Mexico are identified by their tax number (*'Registro Federal de Contribuyentes'*, or RFC), a unique string of alphanumerical characters used to emit and receive invoices, submit tax statements and engage in other procedures. Since 2014, a large number of income and outcome transactions between taxpayers in Mexico have been recorded in CFDIs and stored by SAT.

The data used in this work includes:

- A set of 81,511,015 taxpayer identifiers, anonymized to protect individual identities, which we denote RFCAs, and categorical information for each one of them including: taxpayer type (individual or legal entity), location, date of registration, and economic sector and activity.
- A total of 6,823,415,757 monthly CFDI aggregated emissions between taxpayers distributed between January 2015 and December 2018, which include the RFCA of both the emitter and receiver of the transaction, the month and year, type (either income or outcome), the number of transactions for that month, and the total amounts associated with them.
- A list of 8,570 RFCAs previously identified by Mexican government authorities as definitive or alleged EFOS. We use this data to train machine learning models and as focal points in the network science approach.

In the 48 months of CFDI emissions we analyze, 7,571,093 RFCAs emitted at least one CFDI, so that the already identified evaders (EFOS) account only for 0.0072% of active taxpayers (those who at least emitted one receipt during the period of study). This means that the data is highly unbalanced: the ratio between the identified class (EFOS) and the total number of entities (RFCAs) is quite small, which has an impact in the design of the models and approaches we use.

In what follows, when we refer to a RFCA as an EFOS, we mean those taxpayers already identified by the authorities as either *definitive* or *alleged* evaders. *Unclassified* RFCAs correspond to those that have not been classified as EFOS by the authorities. We denote *suspects* those RFCAs which are classified as possible evaders by our methods, and *nonsuspects* those who don't.

6.3 Results

6.3.1 Deep Neural Networks

As a first classification method of unclassified RFCAs on whether they behave similarly (or not) to definitive EFOS, we implement an *artificial neural network* (ANN). ANNs are models of automatic learning inspired by the human brain. They consist of a collection of interconnected mathematical functions with characteristics analogous to those of biological neurons and are thus called neurons. Just like biological neurons, an artificial neuron collects and classifies information based on its input connections with other neurons, and thereby alternates between an active and inactive state. The connections between neurons have a weight associated with them representative of the intensity of the interaction, such that highly weighted connections are more relevant than those associated with a low weight to modify their activation state. It is via modification of the connection weight between neurons that a neural network learns to identify patterns, a process referred to as *training*.

Neurons of an artificial neural network are often divided into different layers: an input layer that receives data to perform a given task (such as classification); hidden layers that undertake the task through modification of the weights among neurons and adjustment of the input data weights until the task undertaken by the network is optimal; and an output layer, from which the final result of the networks' task over the input data is derived. The output of the network is compared with the desired outcome via a loss function that yields an error quantifier. During training, these errors are propagated through the network to update the weights and minimize the loss function. ANNs have been used in a variety of tasks, including computer vision [5], voice recognition [6], automatic translation [7], board and video games [8–10] and medical diagnostics [11]. They have also been used in a variety of applications in financial services, from forecasting to market studies [12–14], to fraud detection [15] and risk assessment [16, 17]. A neural network can evaluate price data and discover opportunities to make commercial decisions based on data analysis. Networks can distinguish subtle, non-linear inter-dependencies and patterns that other methods of technical analysis cannot. In our case, we use ANNs to perform a classification task and determine whether RFCAs (previously unclassified by the authorities) are actually suspect of illegal tax activity.

6.3.1.1 Data Preparation

In our implementation, we design an ANN that receives input data from all CFDIs associated with an issuing RFCA. The methods used in this section solve a classification task that typically assume a balanced target variable. Since our data contains only 0.0072% already identified EFOS, there is an issue of unbalance that, if not corrected, could skew the results of the algorithm [18]. By means of a technique referred to as *re-sampling* [19], we form a balanced sample of unclassified RFCAs and definitive EFOS. The re-sampling method considered in this implementation is comprised of random re-sampling of the small class (CFDIs issued by definitive EFOS) until it contains as many examples as the other class, in order to finally obtain a large dataset with the same quantity of CFDIs issued by unclassified RFCAs and definitive EFOS.

The model associates each RFCA with a value between 0 and 1 related to the probability that it will be an EFOS. In what follows, we describe the procedure used to design, train and evaluate the ANN. We will follow by presenting some results and conclusions.

6.3.1.2 Modeling

ANN design

A *dynamic recurrent neuronal network* (DRNN) is a special type of neural network that allows introduction of an arbitrary number of rows of data (input variables) at the same time, which is useful in this context since RFCAs have varying amounts of issued CFDIs. Recurrent neural networks are structures in which the output of each execution step provide the input to the following step; this enables them to retain learned information over time. *Long short term memory* (LSTM)[1] describes the design of artificial neurons, i.e., those that give memory to the ANN. These neurons have the best known performance to date and are particularly effective for datasets derived from time series [22–24]. In particular, out of several structures tested, the best performance was obtained with an DRNN with three LSTM cell layers, each with 256 neurons, using a hyperbolic tangential function to calculate their internal state.[2]

An LSTM cell is controlled by three gates: the Forget gate, the Input gate and the Output gate. Each gate within the cell is a different neural network that decides what information is allowed in the cell's state, and which functions as the network's memory. The gates can learn what information is relevant to save or forget during the exercise. The Forget gate controls the amount of information that will be stored in the memory and discards irrelevant information. The Input gate controls the amount of new input that will be stored in the memory, i.e., determines the importance of new information. Finally, the Output gateway determines the characteristics of the analyzed information to obtain an output that will allow correct classification.

The architecture of the neuronal network used to classify RFCA as possible EFOS is made up of three hidden LSTM cell layers, each with 256 neurons connecting a neuron in a layer to a neuron in the following layer. The network unrolls over time to analyze all the invoices issued by an RFCA and, from what has been analyzed, classifies it into EFOS or non-EFOS.

ANN training

To build the ANN, we need a balanced dataset. We start by dividing the dataset of previously identified (definitive) EFOS into two parts: one with 2,981 RFCAs, the *training set*, and one with 745 EFOS, the *test set*. The training set is further extended by duplicating EFOS several times until obtaining a dataset of 1,000,000 RFCAs. Then we add 1,000,000 randomly selected RFCAs from the unclassified dataset, to

---

[1] LSTM cells are a network topology developed for the first time by Hochreiter and Schmidhuber [20] to eliminate the problem of vanishing gradient [21] through the introduction of a memory mechanism. A gradient measures how much the output of a function changes if the input changes a little. The problem is that, for very deep networks, the gradient of errors dissipates rapidly over time, ending up being very small and this prevents a change in the weighted values. Networks with this problem are capable of learning short-term dependencies, but often have difficulty learning long-term dependencies.

[2] These three layers correspond to the hidden layers that classify the input data. In addition to the hidden layers, the network has an input and an output layer.

reach a total of 2,000,000 RFCAs in the training set. The test set is completed by adding 745 random unclassified RFCAs. Notice that both sets will be approximately balanced due to the expected low density of EFOS within the complete dataset. The training set is used to train and adjust the internal parameters of the ANN. Following the training process, the ANN is compared with the test set in order to evaluate its performance.

Additional variables considered

In addition to incorporating the quantitative variables mentioned in Sect. 6.2, we attempted to incorporate categorical data such as the type and situation of the contributor, the situation-description, the status of the contributor, the start date of operations, the sector and federal entity. We also considered incorporating data related to interaction networks (see Sect. 6.3.3), such as the out and indegree, betweenness, closeness, stress, radiality and page rank. However, all the ANN trained with these variables performed equally or worse than the ANN that only used CFDI data.

6.3.1.3 Performance Evaluation

We used the F1-score [25] as a measure to evaluate the competence of the trained model. The F1-score is obtained by calculating the precision harmonic mean and the recall. Precision is the proportion of relevant instances correctly classified out of all the instances that the model believes are relevant. If TP are the true positives and FP the false positives, precision would be given by $TP/(TP + FP)$ (see Table 6.1). Precision answers the question: *How many of the selected RFCAs are actually EFOS?* Recall is the proportion of the incorrectly classified relevant instances out of all the actually relevant instances, $TP/(TP + FN)$, where FN are the false negatives, that answers the question of all RFCAs that are actually EFOS: *How many were correctly classified?* The harmonic mean is defined as the value obtained when the number of values in the dataset is divided by the sum of its reciprocals. It is a type of mean generally used for numbers that represent a ratio or proportion (like the precision and recall) as it equalizes the weight of each datapoint. An F1-score attains its best

Table 6.1 Confusion matrix for binary classification. The true positives (TP) are the examples that the model correctly classified as EFOS. The false negatives (FN) are the examples that the model classified as non-EFOS, but that are actually EFOS. The true negatives (TN) are examples that the model classified as non-EFOS and have not been previously classified as EFOS. The false positives (FP) are examples that the model classified as EFOS, but that were not previously detected as such

		Predicted class	
		P	N
Real class	P	True positives (TP)	False negatives (FN)
	N	False positives (FP)	True negatives (TN)

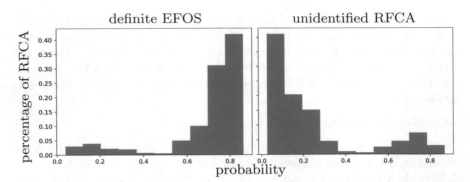

Fig. 6.1 Histograms of the probabilities assigned by the neural network to different sets of RFCAs. (left) We observe that the network correctly assigns most of them a high probability of being EFOS. (right) We observe a bimodal distribution in which there is a considerable percentage of RFCAs that are assigned a high probability of being EFOS

value at 1 (perfect precision and recall) and the worst at 0. Table 6.1 shows a way of separating the classifications made by the neuronal network to allow their evaluation.

For example, if we take 500 definitive EFOS and 500 unclassified ones, and we feed them to our network, we find that TP = 448, FN = 52, TN = 416 and FP = 84. Therefore the precision was 0.845, the recall was 0.896, and we obtained an F1-score of 0.87. If we make the calculation with 1000 alleged EFOS, we obtain $TP = 881$, $FN = 119$ ($TN = FP = 0$ by definition), so the precision is 1, while the recall is 0.881. We thus obtain an F1-score of 0.94.

The RFCAs in the set of "alleged" show the same behavior that the model identified by training with the set of "definitive", and it ends up identifying 88% as EFOS.

In Fig. 6.1 we observe that the model is sure of its decision most of the time (i.e. ends with a very high or very low probability, with a bimodal distribution). Additionally, in the probability distribution of the non-identified RFCAs, there is a percentage that the model is classifying with high probability (i.e., the model is sure that they are EFOS) but has not previously classified them as EFOS.

One of the greatest challenges in neural networks is to interpret what the network is learning from the data. It is not only important to develop a solid solution with great predictive power; it is also of interest to understand the functioning of the developed model, i.e., which variables are the most relevant, the presence of correlations, possible causal relationships, etc. To deepen our understanding of the results, we applied two techniques to determine the more relevant variables that we detail below.

The first technique is based on hypothetical analysis or simulation, and is used to measure the relative influence of the input variables on the model results. In particular, to measure the importance of the variables, we took a sample of our data X and calculated the model predictions, Y, once the model is trained. Then, for each variable x_i we cause a perturbation of this variable (and only this variable) via a normal random distribution centered on 0 with a scale 20% of the variable mean, and calculate the prediction Y_i. We measure the effect exerted by this perturbation

Table 6.2 (Left) Effect of the perturbation on the probability assigned by the neural network; (right) importance of the variables based on the absolute value of the magnitude of the first principal component used to characterize the dataset

Variable	Perturbation effect		Variable	Magnitude
Active subtotal	0.2099		Active total	0.74925125
Active total	0.1813		Active subtotal	0.64598303
Total taxes	0.1419		Total taxes	0.10326791
VAT	0.1083		VAT	0.1032678
Cancelled amount	0.0748		Cancelled amount	0.0000125

by calculating the difference of the quadratic root mean between the original output Y and the perturbed Y_i. A larger mean square root difference means that the variable is "more important". Table 6.2 (left) shows the five variables of greatest importance for the neuronal network.

The second technique consists of the analysis of principal components, a statistical technique to convert high-dimensional to low dimensional data by selecting the most important characteristics that capture most of the information about the dataset. Characteristics are selected as a function of the variation they cause on the output. The characteristic that causes the highest variance is the first component, the characteristic responsible for the second highest variance is considered the second principal component, and so on. It is important to mention that the principal components are not correlated with each other. The importance of each variable is reflected in the magnitude of the corresponding values in vectors that characterize a linear transformation (greater magnitude indicates greater importance). Table 6.2 (right) lists the five variables that best characterize the dataset based on the first principal component, which contributes 99% of the variance. The variable magnitudes are normalized so that the sum of squares is equal to 1.

6.3.1.4   Model Results

The ANN efficiently classifies the identified EFOS that it has been presented with, and by using the trained model, we classify as "suspicious" the unclassified RFCAs to which the ANN assigns a higher probability of behaving similarly to identified EFOS. The ANN classified 149,921 unclassified RFCAs, corresponding to 1.98% of the total, as suspicious, using a threshold probability (> 0.8). The probability threshold to define unclassified RFCAs as suspect can be chosen arbitrarily (0.5 or any other value). We chose the value 0.8 as our threshold to only select those contributors with the highest probability to be similar to the definitive EFOS in our data set, and in this way decrease the number of potential false positives in our classification.

6.3.2 Random Forest

As a second method of classification, we use the automatic technique named *Random Forest* (RF). Techniques of automatic classification, including RF, detect groups of elements with similar statistical patterns in an available dataset, and from the knowledge acquired, make decisions about the membership of new elements in these groups. In our case, we consider the characteristics of EFOS published by SAT and we compare them to unclassified RFCAs.

A RF is constructed by randomly combining different *decision trees* in order to obtain results robust to noise sources inherent to the algorithm. A decision tree is a mathematical algorithm made up of a set of questions ordered and connected to each other through their responses (i.e., the formulation of a question depends on the answer to a preceding question). These questions involve the variables or characteristics of the data utilized. In constructing a decision tree each node represents one of the questions and each fork depends on its answer. Thus, in finishing the construction of a decision tree we can follow a path determined by questions and answers, finally answering the main question: how likely is this RFCA to be an EFOS?

In statistical models like RF it is necessary to maintain a balance between measures such as the *variance* (variability in the prediction of models for different elements) and *bias* (the difference between actual and predicted value). To achieve this balance, an effective technique is the combination of various models, e.g., the combination of decision trees to form a RF. In this way, each decision tree issues a classification (i.e., a probability of suspecting it is an EFOS associated to a RFCA) and the final result of the RF is the most probable classification among all the trees constructed. One of the tasks to resolve in constructing a RF is to find the optimum number of decision trees used to determine the combination that generates the final solution.

In our case, the RF technique is considered adequate since it offers the following advantages:

- Data preparation is minimal. It is only necessary to rely on a dataset where each element to classify, in this case each RFCA, is unique and has a fixed number of characteristics associated with each one of the classes involved, in this case definitive or unclassified.
- It can handle a large number of variables without discriminating any one of them.
- It has been demonstrated that it is one of the methods with highest precision among the classification algorithms [26].
- It performs well with large-volume databases (which applies to the present case study) The result of the RF is a number between 0 and 1 for each evaluated RFCA, which will be interpreted as the probability that each unclassified RFCA is a potential EFOS.

6.3.2.1   Data Preparation

For the implementation of the RF algorithm, the information by issuer is initially grouped, since this analysis is focused on classification of issuing RFCAs. As a result, a unique record for each issuing RFCA is obtained for each of the 48 months considered.

Subsequently, by means of a technique called *undersampling* [27] a balanced sample of unclassified RFCA and definitive EFOS is generated. This technique seeks to determine the optimal number of RFCAs that will allow obtaining a balanced sample of the data (i.e., has the same quantity of unclassified and definitives) as well as a representative one (i.e., that will capture the characteristics of the whole population with the number of RFCAs selected. This process yields a sample with 1561 definitive EFOS and 1561 unclassified RFCAs. The sample obtained so far is the baseline dataset used for implementation of the RF algorithm.

As part of the data preparation phase, two independent treatments are applied to the previously generated sample:

1. The data were analyzed to determine what type of data transformation is viable for each of the sample variables. The family of *box cox* transformations was used to improve the normality of the data and equalize the variance in order to improve the algorithm's performance [28].
2. Principal components analysis (PCA) was used. This consists in reduction of the dimensionality of the dataset by unifying existing variables to create new ones. This method is recommended to improve the performance of the algorithms in question [29].

6.3.2.2   Model Construction

Using the RF algorithm three models were constructed that correspond to the following scenarios and use the sample generated in the previous section:

- First scenario: implementation of the RF algorithm without transformation
- Second scenario: implementation of the RF algorithm using the data sample to which PCA was applied.
- Third scenario: implementation of the RF algorithm using the data sample to which the *box cox* transformation was applied.

For each of the above scenarios, training of the RF algorithm aims to find the optimum number of constituent decision trees. This is achieved by performing iterations of the algorithm, modifying the number of trees used, and determining when the error generated stabilizes at a minimum value. It was concluded that the optimal number of decision tress was 100.

Table 6.3 Comparison of performance measures for the different ways in with the input data were transformed

Scenario	ROC	Error
Random forest	0.912	0.164
Random forest plus principal components	0.886	0.161
Random forest plus variable transformation	0.893	0.157

6.3.2.3 Performance Evaluation

The following measures were used to evaluate the above scenarios:

- Receiver operating characteristic (ROC) curve: is a performance measure with values between 0 and 1; the higher the value, the better the performance is considered. An ROC curve is constructed using the information from two characteristics: the sensitivity (possibility of appropriately classifying a positive individual, in this case a definitive EFOS) and the specificity (possibility of appropriately classifying a negative individual, in this case an unclassified RFCA that is not actually a definitive RFCA) [30].
- Error: is a penalty measure. The closer it is to 0, the better it is considered. The error quantifies the part of the model that is making a mistake in classifying the RFCAs, and in the case of a RF is obtained through a combination of the error generated by each one of the individual trees, as well the correlation between them [26].

As shown in Table 6.3, even though there is an improvement in the performance for the first scenario, error reduction is favored; therefore the model selected was the one that included the Box Cox transformation of data. This was the model used in the following steps.

An additional validation was conducted considering the selected model, which consisted of classifying the definitive EFOS using the model (which we know *a priori* should have have a high probability), and observing the outcome. A cut-off point of 0.8 was established, i.e., if the risk index is greater or equal than 0.8, the classified RFCA is considered an EFOS, otherwise it is not. Additionally, the years of activity of each definitive EFOS were considered for the final diagnosis, e.g., if it was active for two years, the two qualifications are considered, and so on. Results shown in Table 6.4 were obtained by developing the above, where it can be observed that close to 92% of the definitive EFOS are being correctly classified by the algorithm, and the error is only 8%.

Combining the above results, those RFCAs that in all years of activity were detected by the model were classified as possible EFOS, and as non-EFOS if not. Table 6.5 shows that out of all definitive EFOS, only 505 were classified as non-EFOS, which means that they are the only ones about which the algorithm is completely

Table 6.4 We study the performance of the RF algorithm over the years. We considered the definitive EFOS, separated by the number of years with activity (columns). In the different rows, we consider the number of years in which the algorithm classifies RFCAs as EFOS; thus, a definitive EFOS should be detected by the algorithm in at least one of the years of activity. For example, RF erroneously classified 3% out of the total definitive EFOS with activity reported over 3 years, which corresponds to 11 definitive EFOS

	Years with activity			
Years classified as EFOS	1 year	2 years	3 years	4 years
0	17% (133)	5% (56)	3% (11)	6% (8)
1	83% (631)	13% (143)	6% (24)	4% (6)
2		82% (893)	17% (71)	11% (16)
3			74% (307)	26% (37)
4				53% (77)

Table 6.5 Classifying the different types of contributors

Classification	Frequency	Percentage
EFOS	1,908	79%
No EFOS	505	21%

wrong. This behavior is considered normal due to the possibility that the EFOS may have engaged in illegal activities only in some years.

6.3.2.4   Results

Using the model developed and validated in the previous sections (third scenario), we take four groups of unclassified RFCAs (one per year of study) and determine the risk index. Note that if the RFCA has been active for more than one year, it will have a different index each year.

Based on the previous results the following groups were defined for all unclassified RFCAs:

- Suspicious: all those unclassified RFCAs that in each year of activity have a risk index greater or equal than 0.8.
- Not suspicious: all unclassified RFCAs that in at least one of the years of activity have a risk index < 0.8.

Based on these definitions the algorithm classified 7,438,448 RFCAs (98.3%) as not suspicious and 128,227 RFCAs (1.7%) as suspicious of being EFOS.

Fig. 6.2 Directed links in the network correspond to a CFDI emitted between RFCA. This digital receipts (CFDI) can represent incomes and outcomes

6.3.3 Complex Network Approach

In this section we describe the way in which we define interaction networks between EFOS and unclassified RFCA based on the emission and reception of CFDIs. We also describe the analysis we perform on the topology of the network and the roles EFOS and the rest of the RFCA play in them. This analysis allowed us to build the metrics used to define suspect evaders in the interaction networks.

6.3.3.1 Interaction Network Definition

The taxpayers activity records allows us to define interaction networks between them in which nodes correspond to taxpayers (identified by their RFCA) and are classified in one of three categories: definitive EFOS (those evaders already identified by the authorities), alleged EFOS (suspect evaders identified by the authorities), and unclassified RFCA. Links in the network correspond to directed transactions between taxpayers, which as CFDI themselves, can represent either income or outcome transactions (see Fig. 6.2).

The topology of these networks represent the relationships between groups of taxpayers, which we assume reflect some of the association patterns and mechanisms EFOS and other RFCA have used for their practices and which we use to identify suspect evaders.

With the available data we construct yearly and monthly interaction networks. On one hand, the year timescale allow us to identify a set of RFCA with whom EFOS interact more regularly. On the other hand, we have identified on the month timescale, that the amounts associated with transactions between EFOS occur more frequently inside an interval we have termed *EFOS activity regime*, which correspond to higher amounts than those observed in transactions between unclassified RFCA. We build the interactions networks between RFCA, only taking into account the transactions with amounts inside this interval and characterize their topology and structure.

6.3.3.2 Yearly Interaction Networks

We first consider the interaction network built only from the income CFDI emissions and receptions of the nodes associated to EFOS with at least 10 transactions in a year.

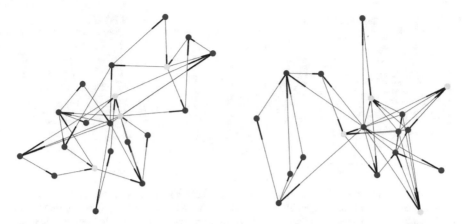

Fig. 6.3 Examples of the strongly connected components observed in interaction networks at year timescale, the left panel corresponds to 2015 and the right panel to 2016. Red nodes correspond to definitive EFOS, yellow to alleged EFOS and blue nodes to unclassified RFCA

This restriction selects the nodes that interact more frequently during a year (at least once a month), which, according to the homophily principle in social networks [31–34], correspond to nodes which are more similar between them.

We identify the strongly connected components (SCC) in these networks with organized sets of taxpayers which can be related to anomalous financial activity. We show the largest SCC observed in 2015 and 2016 in Fig. 6.3. Remembering that links in the network correspond to transactions, the presence of these structures imply a circular flow of goods and services, which being related to nodes associated to EFOS, is possible that these sets of nodes carry out tax evasion practices such as the exchange of receipts of simulated transactions, which suggests that the unclassified RFCA in the SCC might be suspect of carrying out the same practices.

6.3.3.3   Monthly Interaction Networks

In this section we consider the interaction networks between taxpayers at the month timescale. Unlike the yearly interaction networks where we only took into account the emissions and receptions of the nodes associated with EFOS, in this case we consider the transactions between all three kinds of nodes (EFOS, alleged EFOS and unclassified RFCA). Nonetheless, as the whole set of transactions is huge, we need to define criteria to filter transactions to reduce the network into a manageable size.

To this end, we obtain the distribution of the subtotal amounts before taxes of the transactions emitted by EFOS (for both definitive and alleged) to the remaining types of nodes. As can be seen in Fig. 6.4, the median of the distribution changes over time, showing an increase towards the end of the year. It is to be noted that the amounts of the transactions between EFOS are higher than those of the transactions between EFOS and unclassified RFCA, which suggests that EFOS make selective emissions

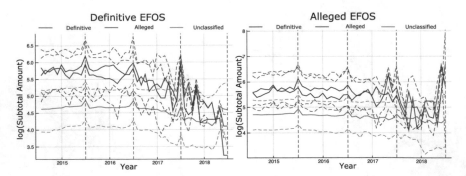

Fig. 6.4 Time behaviour of the logarithm of the subtotal of the transactions associated to emissions from EFOS (left panel) and alleged EFOS (right panel), the vertical dashed lines indicate the month of December of each year. Solid lines show the median and the dashed lines the interquartile range of the distribution. It can be seen that the transactions between EFOS, either definitive or alleged, correspond to amounts around tens of thousands and millions of pesos. We define this range of amounts as the EFOS activity regime, which we use to filter the links in the interaction networks

whether the receiver of the transaction is an EFOS or an arbitrary RFCA. We define the EFOS activity regime, as the amounts interval defined by the interquartile ranges of the amounts distribution obtained from the EFOS transactions. We only take into account links in the monthly interaction networks whose amount lies inside the EFOS activity regime. Once we have selected the relevant links in the monthly interaction networks between taxpayers, we obtain the largest SCC of the network, which e.g. for 2015 consists of 635,588 nodes.

We define the reach of the i-th node in the network at distance d, $r_i(d)$, as the fraction of nodes in the SCC up to a distance d from the node, i.e.

$$\omega_i = \{j \mid d_{ij} \leq d\}, \tag{6.1a}$$

$$r_i(d) = \frac{|\omega_i|}{N}, \tag{6.1b}$$

where, d_{ij} is the length of the shortest path length between nodes i and j, $|\omega_i|$ is the number of elements in ω_i, and N is the number of nodes in the SCC.

If we now consider a set of nodes Ω, we can calculate the mean reach of the set of nodes at distance d, $\langle R(d) \rangle_\Omega$, by

$$\langle R(d) \rangle_\Omega = \frac{1}{|\Omega|} \sum_{i \in \Omega} r_i(d), \tag{6.2}$$

where Ω represents either the set of nodes that correspond to EFOS or unclassified RFCA.

The topology of the network is such that, as can be seen in Fig. 6.5, the reach of EFOS is higher for distances $3 < d < 7$ than for the rest of the nodes. This suggests

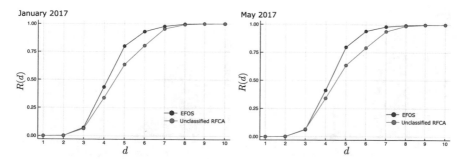

Fig. 6.5 Mean reach, $R(d)$, as a function of distance d, for nodes that correspond to EFOS and to unclassified RFCA. $R(d)$ is higher for EFOS than for unclassified RFCA, which suggests that EFOS are more efficient to distribute their transactions in the network. Data correspond to January 2017 (left panel) and May 2017 (right panel)

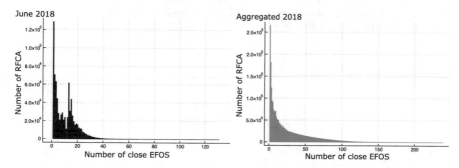

Fig. 6.6 Nearest EFOS to unclassified RFCA distribution for the monthly interaction networks (left panel) and one year's aggregate (right panel). The number of close EFOS is a collusion indicator of unclassified RFCA in the EFOS operation networks. As can be seen in both panels, there are unclassified RFCA close to a large number of EFOS in both timescales

that EFOS are more efficient to distribute their transactions in the network, which can be related to mechanisms aimed to limit the traceability of their transactions. Following the reach behavior, we define the set of nearest neighbors of a node as the set of nodes at $d_{ij} < 3$. We plot the distribution of close EFOS, for both a month and one year's aggregates. From unclassified RFCA in Fig. 6.6, we can see that that there are cases in which unclassified RFCA are close to more than 100 EFOS in one month.

We use the number of close EFOS to unclassified RFCA in the interaction network as an indicator of their collusion in the EFOS operation networks, so that we can assume that an arbitrary RFCA close to a large number of EFOS, is more susceptible to take part in the same corrupt practices as EFOS, when compared to RFCA which are not as close to EFOS in the network.

We define for each one of the identified suspect RFCA, the *EFOS proximity index*, $\sigma_i(y)$, for the i-th node in the network for the year y, as que quotient of the total number of EFOS at $d_{ij} < 3$ during the year, and the number of months these EFOS

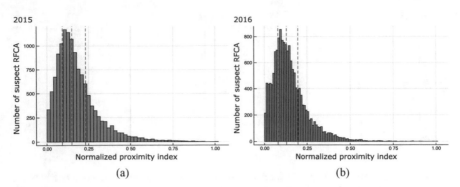

Fig. 6.7 Normalized proximity index to EFOS distribution for suspect RFCA identified by the ANN and Random Forest models. This index is used as an extra validation criteria of the suspect list obtained from the Machine Learning models using characteristics observed in the interaction networks. We show the results for **a** 2015 and **b** 2016. Dashed lines represent the 25, 50 and 75% quartiles of the proximity index distribution

were close to the RFCA, i.e.

$$\sigma_i(y) = \frac{\text{Close EFOS in } y}{\text{Number of months they were close}}. \tag{6.3}$$

Note that the denominator can be less than 12, as there can be months in which the RFCA wasn't close to any of the EFOS in the network. As the number of active EFOS in the network changes over time, we normalize the EFOS proximity index, which we express by $\hat{\sigma}_i$, with respect to the maximum observed during the year, i.e.

$$\hat{\sigma}_i(y) = \frac{\sigma_i(y)}{\max(\sigma_i(y))}, \tag{6.4}$$

where $\hat{\sigma}_i(y) \in [0, 1]$. This allows us to define a threshold to be used in all periods, $\theta_\sigma(y)$, which through the condition $\hat{\sigma}_i(y) \geq \theta_\sigma(y)$ with $\theta_\sigma(y) \approx 1$, allows us to select the more colluded suspect RFCAs for each year (Fig. 6.7).

The description we have made of both yearly and monthly interaction networks between taxpayers, allowed us to identify features of the EFOS organization mechanisms and the local structure of the network around them, such as: their organization in small operation networks related to closed flows of CFDIs between them, and selective CFDI emissions between EFOS of much larger amounts than the transactions they make with unclassified RFCAs, which suggests that EFOS operate inside an activity regime defined by the amounts of their transactions (see Fig. 6.4). We have been able as well to quantify, by means of the reach of nodes in the network and the proximity index, the collusion of unclassified RFCAs inside the EFOS operation networks. These results suggest that complex networks analysis provides useful techniques with ample potential to describe and characterize corruption networks and operational mechanisms, which are yet to be fully explored.

Table 6.6 Number of suspect RFCA identified by each classification method

Method	Classified as suspect	Unclassified
Artificial neural network	149,921	7,416,754
Random forest	128,227	7,438,448

6.3.4 Merging the Classification Methods

We obtain a list of suspect RFCAs from each Machine Learning classification method (ANN and Random Forest). It is to be noted that the training set for both of these methods, consisted of examples from the evaders already identified by the authorities, therefore, there is an inherent and implicit bias in our methods towards the mechanisms and assumptions used by the authorities to identify evaders. This bias in unavoidable due to the data we had available and it is important to take into account further methods to make a wider characterization of these and other evasion mechanisms.

We report the number of suspect RFCAs identified by each method in Table 6.6. The list obtained from the ANN corresponds to those RFCAs with an probability of belonging to the same class as EFOS > 0.8. We use the same threshold for the probability assigned by the Random Forest model. The intersection of both suspect lists is contains 43,650 RFCA, which we consider to have a higher probability of being suspect evaders, as they were identified independently by two different methods.

6.3.4.1 Suspect RFCA Comparison with EFOS

To compare the features of the suspect RFCAs identified by our classification methods with those of EFOS, we compute the distributions of the active and cancelled amounts associated to their CFDI emissions. Active amounts correspond to the emitted and processed invoices, and cancelled amounts correspond to those invoices that were cancelled and not processed. The cancellation of CFDIs can be done freely and independently by taxpayers without an explicit authorization. As can be seen in Fig. 6.8, the distributions of these two variables are very similar for EFOS and the suspect RFCAs identified by our methods, while for nonsuspect RFCA, i.e. those not classified as suspect by our classification methods, the distributions of these variables are different. Furthermore, in Fig. 6.9, we show the time behavior of the active and canceled amounts of the CFDIs emitted by the different RFCA groups (definitive EFOS, alleged EFOS and nonsuspect RFCAs). It can be seen that the difference between suspect and nonsuspect RFCAs is consistent for the whole analysis period, which sets these two groups (EFOS and suspect RFCA) apart from nonsuspect RFCA.

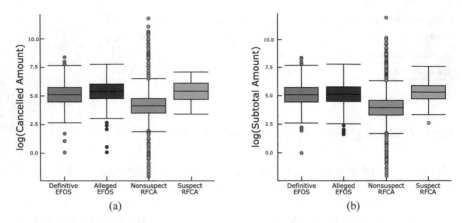

Fig. 6.8 Boxplots for the distributions of the **a** logarithm of the total cancelled amount and **b** the active subtotal amount of the CFDI emitted by EFOS, suspect and nonsuspect RFCA. It can be seen that the distributions for EFOS and suspect RFCA are more similar between each other than when being compared with the distributions of nonsuspect RFCA

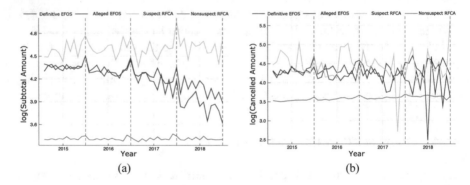

Fig. 6.9 Temporal behavior of the **a** Active subtotal amount and **b** Cancelled total amount for each one of the groups of RFCAs: definitive EFOS (blue), alleged EFOS (red), nonsuspect RFCAs (green) and suspect RFCAs (cyan). The vertical dashed lines correspond to December for each of the years studied. It can be seen that the EFOS and suspect RFCAs behavior differentiates from nonsuspect RFCAs

6.3.4.2 Number of Close EFOS to Suspect RFCA

As it was discussed in Sect. 6.3.3, the EFOS reach in the interaction networks allows us to identify the number of close EFOS to arbitrary RFCA in the network (where close means at a distance $d \leq 3$) and use it to select those RFCAs more immersed in the EFOS operation networks. Now, if we look at the suspect RFCAs identified by the classification methods, and compute the number of close EFOS in a whole year, we observe that they are close to a large number of EFOS (see Fig. 6.10), which suggests that the suspect RFCAs in the intersection of the lists obtained from both classification method are colluded with the EFOS identified by the authorities

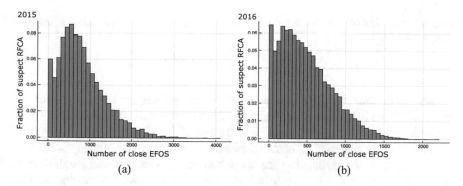

Fig. 6.10 Total of close EFOS the suspect RFCAs identified by our classification methods for **a** 2015 and **b** 2016. It can be seen that a suspect RFCA are close to a large number of EFOS, which indicates that these RFCAs are immersed into the EFOS operation networks

Table 6.7 Type of taxpayer and status of the suspect RFCA identified by the classification methods. Most of them were active legal taxpayers at the moment, which made them susceptible of being investigated by the authorities

Taxpayer type	Percentage of suspect RFCAs
Legal	81.52%
Natural	10.22%
Without info	8.3%

Status	Percentage of suspect RFCAs
Active	91.15%
Cancelled	0.13%
Suspended	0.46%
Without info	8.3%

and gives us confidence about our methods. It is to be noted that the closeness to EFOS wasn't part of the set of features used by the classification methods (ANN and Random Forest) to identify suspect RFCAs, as they were based only on CFDI data.

As we show in Table 6.7, 81.52% of the suspect RFCA are legal persons, which suggests that most part of the emitted CFDI linked with allegedly simulated operations is made between companies or businesses. This can be due to the fact that, in this case the legal responsibility of possible illicit operations, falls to a legal entity and not a natural person. In Table 6.7, we also show that 91.15% of the suspect RFCA were active at the moment of the elaboration of the study, and less than 1% of them had a cancelled or suspended status, which shows that most of the suspect RFCA are economically active, and thus susceptible of being investigated.

Using the data contained in the CFDI, we define the potential tax collection associated to an arbitrary RFCA, $\text{rec}_{\text{VAT}}\phi_i$, as the difference between the total nominal tax obtained by the income CFDI emitted during the year, $\text{VAT}_{\text{Nom}i}$, and the payed tax reported in their tax statements, $\text{VAT}_{\text{Payed}_i}$, i.e.

$$\text{rec}_{\text{VAT}}\phi_i = \sum \text{VAT}_{\text{Nom}i} - \text{VAT}_{\text{Payed}_i}, \tag{6.5}$$

so that, we define the total nominal tax collection that correspond to the set of suspect RFCA, $REC_{IVA}\phi$ as:

$$REC_{VAT}\phi = \sum_i rec_{VAT}\phi_i. \qquad (6.6)$$

We use the income CFDI emitted by the identified suspect RFCA between the years 2015 and 2018, and their yearly tax statements, which include the total VAT payed by them. We believe that the information of the emitted CFDI allows us to have a better description of the economical activity and evasion mechanisms, because the income amounts declared in the tax statements is vulnerable of manipulation, and may not reflect reality, mostly because we are dealing with taxpayers suspects of simulating operations. A significative difference between income amounts from CFDI emissions and tax statements could be a indicator of possible illicit practices.

6.3.5 Yearly Evasion Estimates

As we discussed in Sect. 6.3.3, the number of close EFOS to a RFCA in the interaction networks, is an indicator of their collusion level in the sub-networks associated with EFOS, so that we can assume that a suspect RFCA close to a larger number of definitive and alleged EFOS, is much more susceptible to conduct the same practices as EFOS. With this in mind, we calculate evasion estimates for each year between 2015 and 2018, based on the suspect RFCA closest to EFOS in the monthly interaction networks for each year, which correspond between 28 and 38% of all suspect RFCA.

The EFOS proximity index threshold, θ_σ, allows us to define minimum and maximum values for VAT evasion estimates. The maximum estimated value corresponds to all identified suspect RFCA, and minimum values to estimates based on suspect RFCA of the last quartile (top 25%) of the EFOS proximity index distribution, which correspond to 7,677 unique suspect RFCA with CFDI emissions between 2015 and 2018, which corresponds to an average estimate of 60,604.96 millions of pesos per year. The VAT evasion estimates in Fig. 6.11 should not be considered as final values, as there might exist VAT evasion mechanisms different from the simulation of operations, which we don't take into account in this work. Furthermore, it is important to mention that there is an additional uncertainty because, it is not possible to determine precisely the fraction of simulated and real transactions made by the suspect EFOS with the available information. Because of this situation, we assume the 100% of the emitted CFDI by suspect RFCA as simulated operations. A follow up study focused in the traceability of the emitted CFDI could help determine the fraction of simulated operations, which would allow us to make a more precise VAT evasion estimate.

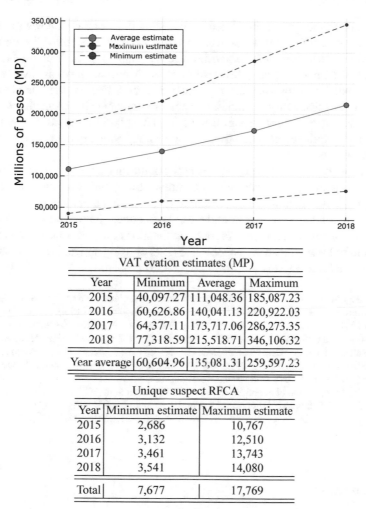

Fig. 6.11 Maximum and minimum VAT evasion estimates in millions of pesos (MP) associated to the emission of CFDI of potentially simulated operations from suspect RFCA in the period 2015 –2018. We present as well the number of unique RFCA used for the calculations for each year. We estimate the number of unique suspect RFCA to be between 7,677 and 17,769

6.4 Discussion

We have shown that it is possible to use tools from network science and machine learning to automatically identify patterns of tax evaders similar to those that humans already identified. This is promising, but should also be taken with caution. It is promising, because similar techniques could be applied in a broad variety of areas: money laundering, bribery practices, and other illegal activities. In principle, this would benefit society. However, one should also be aware of the limits of these methods. First, the patterns identified are based on those already known to humans. This means that different patterns of tax evasion will not be identified. Second, using these techniques to curb illegal activities would promote criminals to use alternative patterns that are not identified, so an arms race would ensue. The tools have a relevant potential, but by themselves, are not a final solution. Finally, there is always the probability of misidentifying honest citizens or companies for wrongdoers, simply because they have similar statistical patterns. Thus, these methods can be used to identify and prioritize *potential* suspects from a huge pool, but the final decision has to be made by humans.

In general, our work illustrates the potential of recent technology to solve different problems exploiting large data sets, computing power, and sophisticated statistical techniques. Still, the limits of this technology are yet to be defined properly, which has led to much hype and overconfidence. Also, ethical issues must be considered for the use of this technology. As more applications are developed, we can learn from them to better situate the benefits and risks of using network science, machine learning, and related techniques to address social problems.

Acknowledgements This manuscript describes research associated with a project advising the Tax Administration Service (SAT) of the Mexican federal government. The official report of the project (in Spanish) is available in the SAT website at http://omawww.sat.gob.mx. We thank Juan Pablo de Botton, Alejandra Cañizares Tello, Leonardo Ignacio Arroyo Trejo, and Aline Jacobo Serrano at SAT, as well as Alejandro Frank Hoeflich, José Luis Mateos Trigos, Juan Claudio Toledo Roy, Ollin Langle, Juan Antonio López Rivera, Eric Solís Montufar, Octavio Zapata Fonseca, Romel Calero, José Luis Gordillo, and Ana Camila Baltar Rodríguez at UNAM. G.I. acknowledges partial support from the Air Force Office of Scientific Research under award number FA8655-20-1-7020, and by the EU H2020 ICT48 project Humane AI Net under contract 952026. C.P. and C.G. acknowledge support by projects CONACyT 285754 and UNAM-PAPIIT IG100518, IG101421, IN107919, and IV100120.

References

1. Cámara de Diputados. Constitución Política de los Estados Unidos Mexicanos. http://www.diputados.gob.mx/LeyesBiblio/ref/cpeum.htm. Accessed 14-04-21
2. Tovar JT (2000) La evasión fiscal: Causas, efectos y soluciones. Porrúa
3. Consultoría SAP. Todo sobre CFDI. https://www.consultoria-sap.com/2018/04/todo-sobre-cfdi.html. Accessed 14-04-21
4. Digital Invoice. Definición EFOS y EDOS. https://digitalinvoice.com.mx/efos-y-edos/. Accessed 14-04-21
5. Hongtao L, Qinchuan Z (2016) Applications of deep convolutional neural network in computer vision. J Data Acquis Process 31(01):1–17
6. Venayagamoorthy GK, Moonasar V, Sandrasegaran K (1998) Voice recognition using neural networks. In: Proceedings of the 1998 South African symposium on communications and signal processing-COMSIG'98 (Cat. No. 98EX214). IEEE, pp 29–32
7. Zhang J, Zong C (2015) Deep neural networks in machine translation: an overview. IEEE Intell Syst 30(05):16–25
8. Tesauro G, Sejnowski TJ (1988) A'neural'network that learns to play backgammon. In: Neural information processing systems (1988), pp 794–803

9. Clark C, Storkey A (2015) Training deep convolutional neural networks to play go. In: International conference on machine learning, pp 1766–1774
10. Sebastian S, He Z, Taku K, Jun S (2019) Neural state machine for character-scene interactions. ACM Trans Grap 38:11
11. Amato F, López A, Peña-Méndez EM, Hampl A, Havel J (2013) Artificial neural networks in medical diagnosis
12. Kimoto T, Asakawa K, Yoda M, Takeoka M (1990) Stock market prediction system with modular neural networks. In: 1990 IJCNN international joint conference on neural networks. IEEE, pp 1–6
13. Mizuno H, Kosaka M, Yajima H, Komoda N (1998) Application of neural network to technical analysis of stock market prediction. Stud Inf Control 7(3):111–120
14. Wilson RL, Sharda R (1994) Bankruptcy prediction using neural networks. Decis Support Syst 11(5):545–557
15. Shen A, Tong R, Deng Y (2007) Application of classification models on credit card fraud detection. In: 2007 international conference on service systems and service management, pp 1–4
16. Trippi RR, Turban E (1992) Neural networks in finance and investing: using artificial intelligence to improve real world performance. McGraw-Hill, Inc
17. Yu L, Wang S, Lai KK (2008) Credit risk assessment with a multistage neural network ensemble learning approach. Expert Syst Appl 34(2):1434–1444
18. Cieslak DA, Chawla NV (2008) Learning decision trees for unbalanced data. In: Daelemans W, Goethals B, Morik K (eds) Machine learning and knowledge discovery in databases. Springer, Berlin, pp 241–256
19. Nathalie Japkowicz. The class imbalance problem: Significance and strategies. In *Proc. of the Int'l Conf. on Artificial Intelligence*, 2000
20. Hochreiter S, Schmidhuber J (1997) Long short-term memory. Neural Comput 9(8):1735–1780 November
21. Hochreiter S (1991) Learning causal models of relational domains. Master's thesis, Institut fur Informatik, Technische Universitat, Munchen
22. Greff K, Srivastava RK, Koutník J, Steunebrink BR, Schmidhuber J (2016) Lstm: a search space odyssey. IEEE Trans Neural Netw Learn Syst 28(10):2222–2232
23. Yin W, Kann K, Yu M, Schütze H (2017) Comparative study of cnn and rnn for natural language processing. arXiv preprint arXiv:1702.01923
24. Chung J, Gulcehre C, Cho K, Bengio Y (2014) Empirical evaluation of gated recurrent neural networks on sequence modeling. arXiv preprint arXiv:1412.3555
25. Van Rijsbergen CJ (1979) Information retrieval, 2nd edn. Butterworth-Heinemann, Newton
26. Breiman L (2001) Random forests. Mach Learn 45(1):5–32
27. Liu X-Y, Wu J, Zhou Z-H (2008) Exploratory undersampling for class-imbalance learning. IEEE Trans Syst, Man, Cybern, Part B (Cybern) 39(2):539–550
28. Osborne JW (2010) Improving your data transformations: applying the box-cox transformation. Pract Assessment, Res Eval 15(12):1–9
29. Wold S, Esbensen K, Geladi P (1987) Principal component analysis. Chemom Intell Lab Syst 2(1–3):37–52
30. Martinez EZ, Neto FL, de Bragança Pereira B (2003) A curva roc para testes diagnósticos. Cadernos de Saúde Coletiva 11(1):7–31
31. Aiello LM, Barrat A, Schifanella R, Cattuto C, Markines B, Menczer F (2012) Friendship prediction and homophily in social media. ACM Trans Web (TWEB) 6(2):9
32. McPherson M, Smith-Lovin L, Cook JM (2001) Birds of a feather: homophily in social networks. Ann Rev Sociol 27(1):415–444
33. Currarini S, Matheson J, Vega-Redondo F (2016) A simple model of homophily in social networks. Eur Econ Rev 90:18–39
34. Aili Asikainen, Gerardo Iñiguez, Javier Ureña-Carrión, Kimmo Kaski, and Mikko Kivelä. Cumulative effects of triadic closure and homophily in social networks. *Science Advances*, 6(19):eaax7310, 2020

Chapter 7
(The Fight Against) Money Laundering: It's All About Networks

Frank Diepenmaat

Abstract In this chapter money laundering will be approached from two perspectives. First, the money laundering concept will be explored. Attention will be paid to the historical background and the functional description of money laundering. Also, the effects of money laundering and the need to tackle this phenomenon will be briefly discussed. From there the focus will shift to the second perspective, the combat of money laundering. The international legal framework for the repression and the prevention of money laundering will be described in general. With that, the existing network in place to fight money laundering will be revealed. Furthermore, attention will be paid to Dutch initiatives to improve the existing network and to create new networks. These are among other aimed at a better cooperation between government bodies and private parties. In time these initiatives might be a valuable contribution to the international fight against money laundering.

7.1 Introduction

In the news money laundering is a recurring topic. Journalists report on mobsters cleaning their proceeds [1], but they also write about banks being involved in money laundering scandals [2]. In this regard, the Danske Bank (Denmark), Pilatus Bank (Malta), the Versobank AS (Estonia), ABLV (Latvia) and ING (the Netherlands) can be named. These money laundering scandals have a large impact on the respective societies and important questions remain. What are the characteristics of the money laundering phenomenon and how can it be tackled? According to several authors the term 'money laundering' was used for the first time in the United States of America during the 1920s [3, 4]. In that period organized crime used laundromat businesses to blur the true origin of its cash. The mafia generated vast amounts of cash with illicit trade, prostitution, gambling but also with the production and sale of alcoholic beverages (the latter prohibited by the National Prohibition Act (Volstead Act) of

F. Diepenmaat (✉)
Saxion University of Applied Sciences, Enschede, Netherlands
e-mail: f.diepenmaat@saxion.nl

1919). To avoid the confiscation of their proceeds, they operated legitimate businesses such as bars, vending machines, hotels, and restaurants. Through these businesses, the illegal money was mixed with the legal proceeds and the total amount was reported as the earnings of the legitimate business. After this process, the money could then be used freely without attracting the attention of law enforcement authorities. Namely, the concept of money laundering depicts a transformation process. In short, the aim of a money launderer is to cover the offences from which proceeds originate—the so-called predicate offences—to use these proceeds for undisturbed consumption or investment.

Unfortunately, there is no ready answer to the question of how money laundering processes are carried out. In fact, money laundering processes take place in secret and they are purely bounded by human creativity [5]. Nevertheless, there are several elements that are essential for any money laundering operation. To explain these elements, it is helpful to describe the money laundering phenomenon from a criminological perspective. In this description, money laundering is depicted as a route consisting out of three stages. It successively concerns the stages of placement, layering and integration. During the placement stage, the proceeds of crime are transferred into the financial system. In most cases, this means that cash money is deposited into a bank account. This stage is aimed at obscuring the direct link between the money and the predicate offence. With that, the money is put into a less suspicious form [6]. The placement stage can be preceded by the transformation of the original proceeds into expensive goods [7]. These goods are sold eventually, and the revenues are then transferred into the financial system. All the actions during—or preceding—the placement stage is aimed at avoiding detection. The perpetrator of the crime tries to distance himself from the predicate offence.

Layering is the second stage of the laundering process. During this stage, the proceeds are sent all over the world, preferably by complex financial transactions. The money in the bank account is wired to several bank accounts in foreign countries, using offshore jurisdictions and shell companies, all to further obscure the link between the proceeds and the underlying offence. In this way, the paper trail, that gives evidence of the true origin of the proceeds, will be veiled until it eventually breaks [3]. Integration is the final stage of the money laundering process. The ill-gotten gains are introduced into the formal economy, for instance using it for consumption or investment [6]. It may be argued that an essential element is lacking in this presentation of facts. The stage of layering and integration coincide without the appearance of legitimacy being attached to the criminal capital. Regardless of which money laundering variant is figured out, each time it is all about breaking the link between the criminal capital and the predicate offence, after which the true origin is replaced by an apparently legitimate origin. In other words, it is the elements of concealment and justification that constitute the nature of money laundering.

If we focus on the element of justification, money launderers simply follow the laws of economics. Money launderers tend to justify their capital in three ways: by presuming the increase of the value of capital, the creation of income or the transfer of capital. In the following, we will investigate these grounds for justification in more depth. Assets that are difficult to valuate in an objective manner, are used to justify crime money. Art, antiques, and real estate can be used for this purpose. By feigning a lucrative sale, a criminal can presume the legal origin of his money. For laundering money in the real estate industry, a market that is not transparent and where it is difficult to estimate values, crime money can be used for the purchase of property first. By selling the property with a vast profit, criminals can create the appearance of a legitimate source and the money can enter the formal economy without suspicion [6].

In economic sectors where large amounts of cash circulate and the turnover is difficult to estimate, income can easily be feigned. Crime money can be justified by presenting it as turnover in bars, hotels, brothels, and casinos. Also, the international trade system can be exploited by money launderers, simply by manipulating trade documents. Trade-based money laundering techniques allow criminals to move large amounts of money undetected under the cover of international trade transactions. By falsely describing goods and services, value movement can be justified. This only requires the misrepresentation of the quality or type of goods or services, such as the shipment of relatively inexpensive goods, which is described as more expensive items [8].

Crime money can also be justified by falsely presenting it as a loan or a donation. In the situation of a loan back-construction, criminal proceeds are placed in the bank account of a foreign legal person. At first glance, there is no relation between this legal person and the criminal, but he is the beneficial owner of this legal entity. From there a loan is given and no suspicion will arise, but in fact, the criminal is just making use of his criminal proceeds [9].

Now we have defined the phenomenon of money laundering, the question is why this phenomenon is at the center of attention in the last decades. The concept—in which the link between criminal capital and the underlying offence is broken, after which the true origin is replaced by an apparently legitimate origin—is from all times. Stolen cattle were branded to conceal the true origin and to justify the presence of the stolen animals in the herd [9]. It was with the rise of organized (drugs) crime that the absolute need for money laundering emerged for the first time. Large amounts of cash money were generated. These amounts turned out to be difficult to hide and these transcended the investments in the informal economy by far. The amount of cash money itself gave indications for its criminal origin [3]. In other words, the rise of organized (drugs) crime led to a relentless generation of profit, which in turn necessitates money laundering.

In parallel with the increased need for money laundering, the opportunities for money laundering increased the last decades. For this, three reasons can be given. First, money launderers benefit from the globalization of the economy and the globalization of the financial system [10]. Enormous amounts of money flow around the globe and between different countries, in which criminal money can escape detection. Secondly, money launderers benefit from technological development. With the introduction of an instantaneous payment system and the internet, the layering stage got an entirely new dimension. As soon as the criminal proceeds are deposited in a bank account, it can be wired endlessly by criminals themselves to obscure the paper trail [11]. Finally, the financial opportunities of criminals must be considered. The generation of vast amounts of profit gives criminals the opportunity to launder their money through professionals. Lawyers, accountants, notaries, and other legal professionals can be hired to act as supervisors or intermediaries to set up and carry out complex money laundering schemes [6]. By involving these facilitators, money laundering is brought to a higher level. Money laundering is carried out under legitimate activities sheltered under client confidentiality. With that, there is evidence that criminals not only benefit from the services of professional money launderers, but also from the services of professional money laundering organizations and professional money laundering networks [12, 13].

7.2   The Necessity to Fight Money Laundering

The incentive for an anti-money laundering strategy was the fight against organized (drugs) crime. As Nadelmann stated, "It was perceived as essential both to identifying and prosecuting the higher-level drug traffickers who rarely if ever came into contact with their illicit goods, and to tracing, seizing and forfeiting their assets" [14]. An effective fight against money laundering could contribute to the preservation of the paper trail, which in turn would make it possible to relate the heads of a criminal enterprise to the predicate offences. So, in this presentation of facts money laundering is a mere side effect of organized crime that needs to be tackled. Over time the insight arose that money laundering is a problem in itself. This was clearly expressed by Vito Tanzi in a 1996 IMF working paper: "The international laundering of money has the potential to impose significant costs on the world economy by (a) harming the effective operations of the national economies and by promoting poorer economic policies, especially in some countries; (b) slowly corrupting the financial market and reducing the public's confidence in the international financial system, thus increasing risks and the instability of that system; and (c) as a consequence (...) reducing the rate of growth of the world economy" [15]. If we focus on the national level, the influx of criminal money in the formal economy might result in the disturbance of economic relations. With that, the risk of reputational damage will threaten entire sectors of the economy if these are related to money laundering operations. And if money laundering is not countered successfully, it might attract more money launderers and it might provoke other types of criminal behavior [16].

   The fight against money laundering should be organized internationally because a national approach merely leads to displacement effects. In that case, criminals will operate their money-laundering operations in those jurisdictions where a comprehensive framework is lacking. Furthermore, the fight against money laundering calls for a two-pronged approach. On the one hand, this approach should include repressive measures such as the confiscation of ill-gotten gains and the criminalization of money laundering. On the other hand, this approach should include preventive measures to protect the formal economy from the influx of the proceeds of crime.

   If we focus on the penalization of money laundering, the functional description presented above, in which it is depicted as a transformation process that consists of at least three successive stages, is of little value. In practice, the different stages can overlap, or they even can occur simultaneously. We just must remind the perpetrator of the predicate offence who is mixing his loot with the revenues of a legitimate business like a hotel. It is with just a single act that the true origin of the money is hidden and replaced by an apparently legitimate origin. So, an effective penalization of money laundering should also apply to acts which in themselves have nothing to do with the disguise and justification of money from crime. Even the accomplice who is performing a minor role throughout—or just before the beginning of the process—must fall within the scope of the criminalization. We must keep in mind the characteristics of the crime. In its nature, the money laundering process is executed in secret. With that, it is a consensual crime since all the people involved are benefiting from it. Lastly, it is considered a victimless crime. Notwithstanding the negative effects described above, money laundering will cause no direct visible harm. An effective criminalization must be tailored to these characteristics. Money laundering should therefore be described in a formal manner in such a way that committing a single act results in a completed offence, so intervention is possible at any time.

If we focus on the prevention of money laundering, the functional description presented above gives valuable insights. This description reveals where and how criminals tend to launder their proceeds and which support is needed in that respect. In the prevention of money laundering, the focus lies on the strengthening of bona fide parties like banks, jewelers, or wholesalers. The aim is to prevent that these representatives of the formal economy turn into facilitators by—consciously or—unconsciously—contributing to money laundering operations. In the following, we will look at the international framework for the prevention of money laundering, and the stakeholders involved, in more detail. At this stage, it is important to note that the repressive—and preventive measures to combat money laundering operate as communicating vessels. The preventive measures must be supported by repressive measures, for those who nevertheless try to integrate their criminal proceeds into the formal economy. On the contrary, the repressive measures must be supported by preventive measures. For the investigation and the prosecution of money laundering cases, in which the paper trail needs to be followed, often information is needed that is obtained because of the execution of preventive measures.

7.3 The Emergence of a Preventive Network

The central idea behind the development of the preventive network against money laundering is that an ounce of prevention is worth a pound of cure. If the link between criminal proceeds and the underlying offence remains intact, criminals can be associated with the crimes committed by them. In this situation, they can be prosecuted and convicted for the predicate offence and their ill-gotten gains can be confiscated. Another important objective in the fight against money laundering is the protection of financial and economic integrity. If money laundering is prevented, the formal economy can be protected against the influx of crime money. As we have seen above, both during- and after a money laundering process, negative effects may arise. It are these effects that pose another argument for the prevention of money laundering. In describing the preventive network, the 2012 Recommendations of the Financial Action Task Force (FATF) [17]—updated in October 2020—will be followed. The FATF is an independent inter-governmental body that develops and promotes policies to protect the global financial system against money laundering, terrorist financing and the financing of proliferation of weapons of mass destruction. The FATF Recommendations are recognized as the global anti-money laundering (AML) and counter-terrorist financing (CFT) standard.

7.3.1 Restrictions on the Transfer of Cash

A first step in the prevention of money laundering is the restriction of the transfer of cash resources. The strength of a preventive network depends on the weakest link. If criminals can transfer their criminal proceeds to foreign countries where no questions are asked, the construction of a preventive network is in vain. The FATF Recommendations [17] require countries to have measures in place to detect the physical cross-border transportation of currency and bearer-negotiable instruments, including through a declaration system and/or disclosure system (FATF Recommendation 32). As an illustration, we turn to the situation within the European Union. There a regulation is applicable that controls cash entering or leaving the EU. In short, carriers who carry cash of a value of 10.000 euros or more are obliged to declare that cash to the competent authorities of the Member State through which they are entering or leaving the Union and make it available to them for control (Article 3(1)). The definition of cash in this instrument is broad. It includes for instance bearer-negotiable instruments and prepaid cards (Article 2(1)(a)). The violation of the rules in this regulation will result in penalties that must be effective, proportionate, and dissuasive (Article 14). With this regulation, a first obstacle is in place. Criminals are discouraged to transfer their proceeds and they are forced to launder their money in an environment where a preventive network is in place.

7.3.2 The Appointment of Gatekeepers

It must be noted that criminals are most vulnerable at the start of the money laundering process. In most situations, the help of third parties is needed to insert crime money into the financial system. A bank employee must deposit large sums of cash on a bank account for instance. During the placement stage, there is still a direct link between the proceeds and the predicate offence. With that, large amounts of cash might give an indication of the criminal origin. The reason for that is that in the formal economy the use of cash has given way to alternative payment methods like online banking and the use of virtual currency. So, it is of the utmost importance to involve private parties in the fight against money laundering. By introducing a set of preventive measures and the appointment of gatekeepers, two goals can be achieved. First, crime money can be kept out of the financial system because criminals are not willing to expose themselves as a perpetrator of predicate offences. Secondly, if they choose to insert their crime money into the financial system nevertheless, a paper trail will arise. This paper trail can support financial investigations at any time in the future. In this way, it remains possible to link criminals to predicate offences and the crime of money laundering.

7.3.3 Obligations for Financial Institutions

Financial institutions must know their customers. It should be prohibited to keep anonymous accounts or accounts in obviously fictitious names. In outline, financial institutions have to undertake customer due diligence (CDD) measures whenever they establish business relations, they carry out occasional transactions above a designated threshold of € 15,000, there is a suspicion of money laundering or terrorist financing or where there is doubt about the veracity or adequacy of previously obtained customer identification data (FATF Recommendation 10 (i-iv). In conducting customer due diligence (CDD) measures, financial institutions must meet four requirements. First, they identify the customer and verify the customer's identity. Secondly, they identify the beneficial owner. For legal persons, this means that the institution understands the ownership and the control structure of the customer. Thirdly, the financial institution understands and obtains information on the purpose and intended nature of the business relationship. Finally, the institution conducts ongoing due diligence on the business relationship and scrutiny of transactions undertaken. Briefly, the institution ensures throughout the course of the relationship that transactions being conducted are consistent with the institution's knowledge of the customer, their business and risk profile and, where necessary, the source of funds [17].

Financial institutions must meet the four requirements set out above. In this, the extent of the measures can be determined using a risk-based approach (FATF Recommendation 1). This means that financial institutions can conduct simplified measures under certain conditions, for instance where they can rely on the information of third parties (FATF Recommendation 17). On the other hand, financial institutions have to apply enhanced measures if a higher risk of money laundering exists. As an example, enhanced measures are required if an institution enters a business relationship with a natural or a legal person from higher-risk countries (FATF Recommendation 19). Where the financial institution is unable to comply with the four requirements, it should be required not to open an account, commence business relations or to perform the requested transaction. Furthermore, it might be required to terminate the business relationship and it should be considered to make a suspicious transactions report in relation to the customer. These requirements apply to both all new customers and existing customers, the latter based on materiality and risk at appropriate times (FATF Recommendation 10).

If a financial institution suspects or has reasonable grounds to suspect that funds are the proceeds of crime, or are related to terrorist financing, it is required to report its suspicions promptly to the financial intelligence unit (hereinafter: FIU) (FATF Recommendation 20). The FIU's main functions are to receive, analyze, and disseminate financial information (FATF Recommendation 29) [17]. Regarding the first function, a distinction can be made between reports that are based on an objective criterion and reports that are based on a subjective criterion. The former is received automatically from a reporting institution whenever a transaction is above a certain threshold. For instance, if a financial institution receives a cash deposit that amounts to € 15,000 or more, it is obliged to report this to the FIU. As for the latter, a report is based on the judgement of the reporting institution that the transaction might be related to money laundering or terrorist financing. This can be the case when the transaction does not fit within the nature of the business relationship, or when an economic rationale for the transaction seems to lack. Furthermore, the FIU can receive information from domestic agencies responsible for supervision as well as from foreign FIU's. As to the latter, reference must be made to the Egmont Group, a platform for the secure exchange of expertise and financial intelligence between 166 FIU's. The second function of FIU is to analyze the reports received from the reporting entities. The purpose of this analysis is to establish whether the data contained in the reports provide sufficient evidence related to money laundering or terrorist financing and whether the data could be used for further investigation and prosecution. The third function concerns the dissemination of financial intelligence to competent authorities. Here the FIU can transmit information to the law enforcement agencies if the analysis revealed money laundering or other types of crime. With that, information can be shared with anti-money laundering supervisors and foreign FIU's.

If we focus on the position of the FIU, this position is in the middle between the financial sector and law enforcement agencies, which are responsible for the investigation and prosecution of crime. The FIU first processes suspicious transactions as reported by the financial institutions, before disseminating information to the police and public prosecution. Because of this, financial institutions do not have to be reluctant to report findings regarding suspicious transactions. The reported information is thoroughly analyzed and only the transactions revealing criminal elements will be forwarded to the law enforcement agencies. From there money launderers can be prosecuted and the proceeds of crime can be seized and eventually confiscated.

7.3.4   Obligations for Designated Non-financial Businesses and Professions

The preventive network against money laundering and the financing of terrorism increased over the last decades. More gatekeepers have been appointed since criminals found new ways to launder their money. The introduction of obligations for financial institutions in the fight against money laundering simply caused displacement effects. A few examples to illustrate this. If no measures are taken, criminals can resort to casinos to place their ill-gotten gains. In a casino, cash will be converted into chips. From there many chips are brought together and these are offered to the cashier, feigning that these represent winnings. The chips are converted, and the cashier is kindly asked to wire the money into a designated bank account. Criminals might also circumvent preventive measures by converting cash into valuable goods like jewelry. These can be sold or transferred to foreign countries easily, so the link with the predicate offence slowly disappears. Furthermore, the insight arose that criminals use the services of lawyers, notaries, and accountants for money laundering purposes. During the layering stage of the process, criminals try to obscure the true origin of the proceeds. They also try to obscure ownership and control. In its development as a complicated and sophisticated crime, money laundering might be conducted through or under the cover of corporate entities. By using legal entities, criminals can stay out of the picture. They can invest in real estate for example, without raising any suspicion. For the outside world, someone is a tenant. This person is the owner since he pulls the strings within the legal entity that bought the real estate. The common denominator in these examples is that representatives of the formal economy—consciously or unconsciously—are involved in money laundering processes. The preventive network against money laundering has therefore been expanded. The obligations regarding customer due diligence and the reporting of a suspicious transaction as discussed above, are in the meanwhile to the utmost extent applicable to a wide range of non-financial businesses and professionals (FATF Recommendations 22–23). In this way, the pass of money launderers can be cut, and the financial and economic integrity can be assured.

7.3.5 Regulation and Supervision of Gatekeepers

An important presupposition exists in the appointment of private parties as gatekeepers in the fight against money laundering and terrorist financing. This presupposition is that these are bona fide parties, who are willing to protect the financial and economic integrity by preventing that crime money reaches the formal economy unnoticed. If criminals can occupy crucial positions in the formal economy themselves, they can easily circumvent the existing measures. This calls for regulation. For financial institutions, like banks, this means that legal or regulatory measures must be in place to prevent criminals or their associates from holding, or being the beneficial owner of, a significant or controlling interest, or holding a management function in, a financial institution (FATF Recommendation 26). The same applies roughly for designated non-financial institutions and professions (FATF Recommendation 28). A licensing system makes it possible to keep criminals out of the financial sector. In most countries, it is only allowed to provide financial services based on a license issued by the Central Bank. For the Dutch situation, we can point for instance at the Financial Supervision Act (Wet op het Financieel Toezicht/WFT). With that, in some countries, a screening system exists for sectors that are vulnerable to the influx of criminal proceeds and the committing of crimes. In these countries for instance a screening is required to open a bar, a hotel, or a brothel. An example is the Dutch Public Administration (Probity Screening) Act (Wet bevordering integriteitsbeoordelingen door het openbaar bestuur/ Wet Bibob). In this way, it can be prevented that known criminals could integrate crime money into the cash flow of legitimate businesses. It should be kept in mind that for the purpose of money laundering, they only must put criminal proceeds in the cash register, later stating that it is turnover. In short, barriers are created to prevent that criminals can launder their proceeds themselves. They remain dependent on gatekeepers who are included in the preventive network against money laundering and terrorist financing.

From there the supervision of the behavior of the gatekeepers is important. The preventive measures, including customer due diligence and the reporting of suspicious transactions, are not effective if gatekeepers turn a blind eye. Therefore, national authorities must supervise or monitor the designated gatekeepers for compliance with these preventive measures (FATF Recommendations 26–28). The national supervisors have the authority to conduct inspections and to impose sanctions for failure to comply with the applicable requirements. These sanctions include a range of disciplinary and financial sanctions. These might even include the power to withdraw, restrict or suspend the license of the institution. The supervisory task can also be performed by a self-regulatory body, provided that such a body can ensure that its members comply with their obligations to combat money laundering and terrorist financing. As we have seen above, a wide range of gatekeepers is in place. The supervision of these gatekeepers is arranged likewise. If we take the Dutch situation as an example, financial institutions like banks, credit institutions, investment institutions and financial service providers are supervised by the Dutch Central Bank (DNB) and the Dutch Authority for the Financial Markets (AFM). The Dutch Tax Authority/Wwft Supervision Office is the supervisor for the non-financial institutions such as real estate agents, traders and sellers of goods and pawnshops. The Financial Supervision Office (BFT) and the local Dean of the Bar Association are the supervisors for professions such as accountants, tax advisers, notaries, and lawyers. Supervisors have an important role to ensure compliance by the gatekeepers. With that, they have the duty to report suspicious transactions to the FIU if they observe that a gatekeeper failed to do so. Criminals that were benefiting from the fact that a gatekeeper failed to comply with the measures on customer due diligence and the reporting of suspicious transactions, will not go free. The relevant information will be processed by the FIU and eventually brought to the attention of the law enforcement agencies. In the end, this can result in a conviction for money laundering and the confiscation of the ill-gotten gains. It is with the involvement of the supervising agencies, that the preventive network is complete (Fig. 7.1).

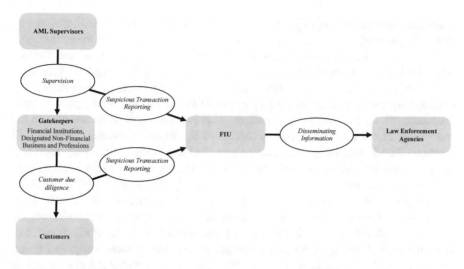

Fig. 7.1 The existing preventive network

7.4 The Improvement of Existing Networks and the Development of New Networks

Notwithstanding the preventive network in place, the amount of money laundered is still mind-blowing. The United Nations Office on Drugs and Crime estimates that the amount of money laundered globally in one year is two to five per cent of the global GDP or $800 billion to $2 trillion in current US dollars [18]. Unger et al. [19] come to an estimate of 1,2% of the global GDP, a total of 677 billion euro. For the Netherlands, they estimate that the amount of money laundered in 2014 equals 2,5% of the Dutch GDP or a total of 16 billion euro [19]. At the same time, the effectiveness of the preventive measures to combat money laundering is questioned from time to time. The obligations regarding customer due diligence and the reporting of suspicious transactions place a burden on the appointed gatekeepers, but from there—and after the analysis of FIU-the Netherlands—the information disseminated plays a minor role in further investigations [20, 21]. This calls for a change. A change on the side of the law enforcement, including the increase of priority, additional funding, and the improvement of expertise to fight money laundering more effectively, but that will not be our perspective in the following. We will focus on the improvement of the preventive network already in place. Furthermore, we will focus on the development of new networks in which available information is shared between relevant stakeholders. We can use the metaphor of a puzzle to clarify this. Private parties, the public administration and law enforcement agencies all have access to information that can be presented as the individual pieces of a puzzle. Separate from each other, they have no overview. By sharing information, so by putting the pieces of the puzzle in place together, an image occurs. Networks in which available information is shared between the participating parties can both contribute to the revealing of money laundering operations and to the protection of vulnerable sectors of the economy.

The establishment of Transaction Monitoring Netherlands (hereinafter: TMNL) will give impetus to the customer due diligence of banks [22]. Five Dutch banks—ABN AMRO, ING, Rabobank, Triodos Bank and the Volksbank—have decided to establish Transaction Monitoring Netherlands (TMNL) in the fight against money laundering and the financing of terrorism. This initiative will be an addition to the individual transaction monitoring activities of the banks mentioned. TMNL will focus on the identification of unusual patterns in the payment traffic that individual banks cannot identify. As we have seen above, criminals make every effort to conceal the origin of their criminal funds, and frequently they abuse multiple banks for this purpose. Effective combat of money laundering is only possible through closer cooperation. The combination of the transaction data of several banks will provide new inter-bank information. In this way, TMNL will allow for more effective detection of criminal money flows and networks in addition to what individual banks can achieve with their own transaction data. The banks are working close together with government partners such as the Ministries of Finance and Justice and Security and the FIU. In this, they are not only fulfilling their responsibility as gatekeepers. In addition, banks become part of a national chain that aims at the identification, detection, prosecution, and conviction of money launderers. In this way, banks are not only preserving the integrity of the financial system. They are also contributing to the prevention and the repression of the undermining influence of crime. In due course, other banks will also be able to make use of TMNL. Furthermore, the banks' individual transaction monitoring activities might be outsourced to TMNL in the near future, if this organization proves to function effectively.

Another initiative worth mentioning is the Fintell Alliance. This concerns a public-private partnership between FIU-the Netherlands and several banks. This partnership is aimed at exchanging financial intelligence and strengthening the effectiveness of the reporting of suspicious transactions. This cooperation results in a better insight into criminal networks, facilitators, and the laundering of criminal assets (FIU-the Netherlands 2019: 19). This cooperation also offers a valuable feedback loop, which may result in the improvement of the quality of the issued reports. From there, this may contribute to the effectiveness of the reporting of suspicious transactions. Besides the improvement of the preventive network in place, by focusing on customer due diligence and the reporting of suspicious transactions, new networks appear. The Anti Money Laundering Centre (hereinafter: AMLC) is a platform where the parties involved in combating money laundering can share their knowledge and expertise and where they can work together operationally (AMLC). Affiliated parties are among other the Fiscal Information and Investigation Service (FIOD), the police, the Public Prosecution Service, FIU-the Netherlands, banks, exchange services for virtual currency and universities. This public-private partnership is primarily aimed at an increase of the level of knowledge, analyzing trends and identifying new money laundering methods. Within the Financial Expertise Centre (hereinafter: FEC) various public partners like the Dutch Central Bank, the Tax and Customs Administration, FIOD, FIU-the Netherlands, the police, the Public Prosecution Service, and

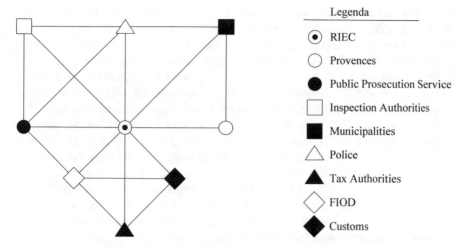

Fig. 7.2 The network of the RIEC

the Dutch Authority for the Financial Markets (AFM) work together to strengthen the integrity of the financial sector by promoting cooperation and the exchange of information. Within this network, also the opportunity for cooperation with private parties exists [23]. This public-private partnership, with the name FEC PPS, resulted in a joint project on trade-based money laundering, in which for instance a particular sector—the automotive sector—was examined and training material for banks was developed.

The main goal of the Regional Intelligence and Expertise Centre (hereinafter: RIEC) is to fight subversive crime ('ondermijning'). On the one hand, the intelligence and the expertise of public parties are connected within this network. On the other hand, it stimulates and supports public-private partnerships in the fight against subversive crime (RIEC). Since money laundering is considered a subversive crime, attention is paid to this phenomenon within the ten RIEC's that are present in the Netherlands. A broad range of parties are affiliated, such as municipalities, provinces, the police, the Public Prosecution Service, the tax authorities, FIOD, customs and Inspection authorities. By sharing information, money laundering processes can be revealed. If the activities in a particular neighborhood are analyzed together, anomalies can be noticed which easily would evade detection. The restaurant owner can present criminal proceeds as legitimate turnover without raising suspicion from the tax authorities. This alters when representatives of the municipality or the police state that this restaurant is only open during weekends (Fig. 7.2).

The infobox for criminal and unexplained assets (hereinafter: iCOV) is a collaboration between several government organizations, including among other FIU-the Netherlands, the Dutch Central Bank, the tax authorities, customs, the police and the Public Prosecution Service. This entity provides different types of data intelligence reports that help the affiliated organizations to effectively intervene [24].

The iCOV capital and income report (iRVI) gives a full overview of a person's or company's financial situation, including the history. It shows for instance all details about income, ownership of the real estate, bank accounts, debts, and inheritance. Since iCOV has access to this data under strict conditions, financial reports can be generated without the need to approach each organization separately. These capital and income reports can help to tackle money laundering since they can reveal discrepancies between a factual financial situation and an actual financial situation. The iCOV relation scan (iRR) shows all formal relations of a person. This will show the relations with family and business relations, but also the relations with criminal contacts. In short, this relation scan is a customized process in which the network around a person can be described in detail. This may bring money laundering constructions to light and it can give directions for further investigation. Finally, the iCOV report theme (iRT) concerns reports, on money laundering through real estate in a specific geographic area. For instance, a report can be generated on the financing of real estate in a particular street. If mortgages are granted by companies in Panama or from the British Virgin Islands, that may be a starting point for further investigation since it might indicate a loan-back construction. In sum, the combination of hard data and human interpretation offers valuable insights to tackle money laundering and to confiscate criminal or unexplained assets more effectively.

7.5   Future Challenges

It shall be no surprise that money launderers always will try to find new ways to break the link between the criminal capital and the predicate offence, to replace the true origin with an apparently legitimate origin. New methods will be invented, and new networks will arise. This is inherent to the nature of this crime. An effective fight against money laundering requires a response to this. Since money laundering is a threat to the financial- and economic integrity and the soundness of the formal economy, it is a concern to all. An effective fight against money laundering therefore not only involves law enforcement agencies. It also involves the public administration and private parties. Preventive networks in place must be adaptive and new networks—with a variety of perspectives—must be formed in order to cut the pass of money launderers. Close cooperation between all stakeholders is crucial and information must be shared generously. The functional description of money laundering remains helpful to determine the right perspective: which sectors of the economy are vulnerable for the integration of criminal money; how can criminal proceeds be justified there and who might be involved? To create a maximum impact, the focus must be drawn to professional money launderers here [25]. The concepts and tools of complex networks science can be used to signal money laundering and to assemble evidence [26, 27]. In sum, an effective fight against money laundering requires persistence and creativity at all levels of society. In fact, it is all about networks in which all the pieces of the puzzle are put in place together.

References

1. Humphreys A, The underworld laundromat: how to clean 10 million in mob money. https://nationalpost.com/news/theunderworldlaundromathowtoclean10millioninmob-money. Accessed 31/01/21
2. Jensen T, Explainer: Danske bank's 200 billion euro money laundering scandal. https://www.reuters.com/article/danskebankmoneylaunderingidINKCN1NO13V. Accessed 31/01/21
3. Stessens G (1997) De nationale en internationale bestrijding van het witwassen. Intersentia, Antwerp
4. Hinterseer K (2004) Criminal Finance. Kluwer Law International, The Hague
5. Gilmore W (2004) Dirty money. Council of Europe Publishing, Strasbourg
6. Amrani H (2012) The Development of Anti-Money Regime: Challenging issues to sovereignty, jurisdiction, law enforcement, and their implications on the effectiveness in countering money laundering. Erasmus University Rotterdam, Rotterdam
7. Akse T (2003) En de kleur is vuil. Zoetermeer: Korps Landelijke Politiediensten. Dienst Nationale Recherche Informatie, Amsterdam
8. [FAFT]. Trade-based money laundering: trends and developments. https://fatf-gafi.org/publications/methodandtrends/documents/trade-based-money-laundering-trends-and-developments.html. Accessed 31/01/21
9. Diepenmaat F (2016) De Nederlandse strafbaarstelling van witwassen. Een onderzoek naar de reikwijdte en de toepassing van artikel 420bis Sr. Kluwer, Deventer
10. Morris-Cotterill N (2001) Money laundering. Foreign Policy 124(1):16–22
11. Met-Domestici A (2013) The reform of the fight against money laundering in the eu. Eucrim 3:89–95
12. [FAFT]. Professional money laundering. www.fatf-gafi.org/publications/methodandtrends/documents/professional-money-laundering.html. Accessed 31/01/21
13. Kramer JA, Blokland A, Soudijn M (2020) de verborgen netwerken van witwassers. Witwassen als bedrijfsmatige activiteit. Tijdschrift voor Criminologie 4:365–382
14. Nadelmann E (1994) Cops across borders: the internationalization of US criminal law enforcement. Penn State University Press, University Park, PA
15. Tanzi V (1996) Money laundering and the international financial system. IMF Work Pap 1996(055)
16. Unger B, Rawlings G, Siegel M, Ferwerda J, de Kruijf W, Busuioic M, Wokke K (2006) The amounts and the effects of money laundering. Utrecht School of Economics, Utrecht
17. [FAFT]. International standards on combating money laundering and the financing of terrorism and proliferation. https://fatf-gafi.org/recommendations.html. Accessed 31/01/21
18. [UNODC]. Money laundering. https://www.unodc.org/unodc/en/money-laundering/overview.html. Accessed 31/01/21
19. Unger B, Ferwerda J, Koetsier I, Gjoleka J, van Saase A, Slot B, de Swart L (2018) Aard en omvang van criminele bestedingen. Utrecht School of Economics-Ecoris, Utrecht/Rotterdam
20. [Dutch Court of Audit]. Bestrijden witwassen en terrorismefinanciering. https://www.parlementairemonitor.nl/9353000/1/j9vvij5epmj1ey0/vhvhhhh2szzd. Accessed 31/01/21
21. [FAFT]. Mutual evaluation report the netherlands. http://www.fatf-gafi.org/media/fatf/documents/reports/mer/. Accessed 31/01/21
22. [Dutch Banking Association]. Transaction monitoring netherlands: a unique step in the fight against money laundering and the financing of terrorism. https://www.nvb.nl/english/transactionmonitoringnetherlandsauniquestep-in-the-fight-against-money-laundering-and-the-financing-of-terrorism/. Accessed 31/01/21
23. [FEC]. Annual plan 2020. https://www.fec-partners.nl/nl/publicaties/fec_jaarplan. Accessed 31/01/21
24. [iCOV]. English explanimation. https://icov.nl. Accessed 31/01/21
25. McCarthy KJ, van Santen P, Fiedler I (2015) Modeling the money launderer: Microtheoretical arguments on anti-money laundering policy. Int Rev Law Econ 43:148–155

26. Garcia-Bedoya O, Granados O, Burgos JC (2020) AI against money laundering networks: the Colombian case. J Money Laund Control. https://doi.org/10.1108/JMLC-04-2020-0033
27. Luna-Pla I, Nicolás-Carlock JR (2020) Corruption and complexity: a scientific framework for the analysis of corruption networks. Appl Netw Sci 5(1):13

Chapter 8
Financial Networks and Structure of Global Financial Crime

Oscar M. Granados and Andrés Vargas

Abstract Financial crimes are social problems that affect many different communities around the world, involving public and private organizations in diverse sectors and activities. We analyze global financial networks focusing on particular aspects of their characteristics when suspicious activities of tax fraud, corruption, and money laundering could be identified. This research provides a perspective on the presence of financial crime phenomena in global financial networks through the study of their large-scale structure. Our results reveal that suspicious activities run in small groups, and they emerge around communities of financial intermediaries, non-financial intermediaries, and offshore entities. Moreover, we find preliminary indications that these activities can play a role in deviating the degree distributions of these networks from a power-law behavior. We also discuss the temporal evolution of the networks and its significance in the identification of suspicious activities and financial crime.

8.1 Introduction

Different reports from the International Monetary Fund (IMF) and the United Nations Office on Drugs and Crime (UNODC) estimate that the amount of money laundered around the world in one year is between 2 and 5% of the global Gross Domestic Product (GDP) [1]. Some part of this money comes from tax fraud, corruption, and another part is a result of funds generated by the drug, human, and arms trafficking and organized crime. But the exact amount of money is impossible to estimate due precisely to the undercover nature of the crime.

O. M. Granados (✉)
Department of Economics and International Trade, Universidad Jorge Tadeo Lozano,
Carrera 4 22-61, Bogotá, Colombia
e-mail: oscarm.granadose@utadeo.edu.co

A. Vargas
Departamento de Matemáticas, Pontificia Universidad Javeriana, Carrera 7 40-62,
Bogotá, Colombia
e-mail: a.vargasd@javeriana.edu.co

Financial crimes are global phenomena that transcend economic, cultural, and social borders of developed and developing countries, especially when tax evaders, corrupted individuals, and criminals use the global financial system to protect their illegal money. Since the success of tax fraud and corruption lies frequently in the transformation of illegal into legal money through the money laundering process, separating these crimes is a complex task that is more difficult when these activities use the financial system to transfer funds between different jurisdictions. But by this same characteristic, money laundering can be an indicator of corruption and tax fraud because of the actions related to financial crimes appear the movements to keep off money from their geographic or financial source through the trading of financial instruments, or wired through different banking accounts anywhere around the world as tax havens or other jurisdictions. They expect that illegal money will return to them after several transactions.

It has been common to study these phenomena separately and from different perspectives, using particular methodologies and criteria according to the chosen approach. Indeed, previous studies have investigated corruption as a country scale problem [2–10], as a result of a broad spectrum of government interactions [11–18], as a result from links with private interests [19–24] or criminal activities [25, 26], to name just a few among many other examples. In turn, money laundering has been analyzed through different approaches using an econometric model [27], an economic growth model [28], global economics methods [29, 30], microeconomics frameworks [31–33], a gravity model [34], social network analysis [35–37], or network visualization with artificial intelligence methods [38], among others.

On the other hand, the complexity of regulations and financial instruments creates differences across jurisdictions that can be exploited by firms for tax minimization purposes and other motivations. This is the case, e.g., with tax arbitrage interests of multinational companies, financial institutions, investment and wealth managers, or family offices. In this way, financial markets and tax havens become the best options to arbitrage opportunity between high tax and low tax jurisdictions [39–41]. As a consequence, in the last decades, several global financial institutions have created products and services, as well as correspondent structures and branches in these jurisdictions for their global clients. However, several governments and international organizations criticize financial services in low tax jurisdictions because they see them as activities that promote tax fraud, corruption, and money laundering. In fact, some scholars have concluded that tax havens provide low regulatory standards favorable for illegal activities [42], and the intensive use of certain financial transactions in tax havens is held as suspicious. Examples of these activities are outbound deposits, understood as the transference of money from a non-haven jurisdiction to a tax haven jurisdiction, or resorting to deposit triangulation between the same agents [43].

We note that several studies consider particular types of financial crime separately. However, tax fraud, corruption, and money laundering are complex and highly interrelated crimes that take advantage of financial networks around the world, especially those located in tax havens, and an important task that motivates our study is to understand how these activities are reflected in the financial networks structure. To analyze this situation, we explore financial network data sets, using information available from the Bahamas Leaks, the Panama Papers, and several related sources. These data sets provide details of interactions among thousands of agents as financial intermediaries, non-financial intermediaries, and offshore entities in different jurisdictions. Through them, we obtain insights into important characteristics of the financial dealings that are part of the backdrop of global financial crimes.

Applying statistical techniques, we elucidate how close the large scale structure of these financial networks reflects a scale-free behavior (as it has already been observed for some economic networks [44]). Moreover, we analyze their dynamics using network science methods to understand, when possible, their evolution, and to identify suspicious interactions. We find visible agents and interaction facilitators as well as many other agents that go unnoticed in these networks and whose activities develop in different ways. For example, suspicious groups evolve around some financial institutions, non-financial intermediaries, and offshore entities that facilitate the different kinds of financial crimes. In addition to reveal partially illegal activities, our results indicate new perspectives to study problems related to financial crime in large-scale networks.

This paper is organized as follows. In Sect. 8.1, we present some basic mathematical and financial terminology. Section 8.2 describes the data sets and presents a brief description of communities and financial network structure methods. Section 8.3 contains the jurisdiction and network analyses, the results of community detection and explains the dynamical structure of the largest suspicious activities. The paper closes with Sect. 8.4 where we highlight important aspects of our approach and provide directions for future work.

8.2   Basic Notions

We provide some definitions of the fundamental mathematical and financial terminology required in our analysis. For more details, we refer the reader to [45, 46].

8.2.1   Network Definitions

⋄ A *graph* $G = (V, E)$ is a pair consisting of a set V whose elements are called *nodes* or *vertices*, and a set E of pairs of nodes called *edges*. A sub-graph $G' = (V', E')$ of G is a graph with $V' \subset V$ and $E' \subset E$, where E' contains all the elements of E connecting vertices of V'.
⋄ A *path* is a finite sequence of distinct edges joining a sequence of distinct nodes.
⋄ A *shortest path* is a path of minimal length between two nodes.
⋄ The *degree* of a vertex $v \in V$ is the number of nodes to which it is connected.
⋄ A *centrality measure* is a way to quantify the importance of a vertex of G with respect to some particular criteria. The degree can be used as an example.

◇ *Assortativity* is a measure of the preference that the nodes of G have to attach to others that are similar in certain way. The positive values of assortativity indicate a preference for connections between nodes of similar degree, while negative values indicate a preference for connections between nodes of different degree.

8.2.2 Financial Terminology

◇ A *jurisdiction* is a geographical location subordinated to some judicial, law, financial, or other enforcement authority and where financial or business activities take place.
◇ A *tax haven* is a jurisdiction with low rates of taxation for foreign investors.
◇ An *intermediary* is an institution that connects financial and business operations and its roles include a broad range of activities like financial, consulting, legal, accounting, and management services.
◇ An *offshore entity* is a company incorporated in jurisdictions with low rates of taxation (tax havens) and its activities are developed overseas, i.e., the company does not undertake business with persons resident in that jurisdiction. Additionally, its profits are not repatriated.
◇ An *officer* is a person who represents legally an intermediary or an offshore entity.

8.3 Methods

8.3.1 Data Collection

The financial networks that constitute our object of study were constructed from the Bahamas Leaks and Panama Papers data sets based on the records of the International Consortium of Investigative Journalists [47]. These data sets contain information about 0.7 million agents (intermediaries, offshore entities, and officers) and almost 1 million edges among these agents, especially, interactions of how intermediaries and offshore firms used unobservable instruments to develop their activities. We have used the same data structure of the source where the data set is divided in different subsets with information of intermediaries, offshore entities, officers, and one additional subset with the matrix of interactions. For legal concerns, we did not use names or other information that could be sensitive.

The Bahamas Leaks came from its national registries and almost all offshore entities and intermediaries had jurisdiction in the Bahamas, i.e., this data set is a result of country-centric leaks. The Panama Papers are a result of a leak of confidential documents concerning Mossack Fonseca activities, where the offshore entities

Table 8.1 Number and type of agents and interactions

	Role	Type	Sub-type	Bahamas leaks	Panama papers
Agents	Intermediaries	Financial	Funds Banks Private banks Trust companies Wealth managers	541	14,110
		Non-Financial	Legal Consulting Accounting Management		
	Offshore entities			175,888	213,634
	Officers			25,262	238,402
Edges	Interactions	Financial Legal Consulting Accounting Management		246,291	657,488

had jurisdiction in different places around the world, but especially in Caribbean jurisdictions. We compiled and curated the data using the following criteria:

⋄ The data sets contain historical aggregate information for the period 1980–2017 with unweighted values and undirected interactions.
⋄ We did not use the data containing non-classified agents.
⋄ We did not use data concerning the names of agents and information sources.
⋄ We removed all data about agents without jurisdiction or non-complete information.

Using these data, we constructed two networks whose nodes correspond to agents of three roles (see Table 8.1): intermediaries, offshore entities, and officers. Intermediaries are divided into two types, financial and non-financial. The financial intermediaries comprise all financial institutions like banks, private banks, trust companies, funds or wealth managers with operations in the Caribbean jurisdictions, and the non-financial intermediaries refer to companies of consulting, legal, accounting, and management services for corporate and private entities. Offshore entities are companies operating outside of the jurisdiction where they were originally established. Officers are persons needed to represent company activities and, habitually, are employees of non-financial intermediaries. In some cases, they are employees of financial intermediaries.

Furthermore, the agents in the networks (intermediaries, offshore entities, and officers) may be connected between themselves through five types of interactions:

financial, business, legal, accounting, and management relations (see Table 8.1). Any of these interactions determine an edge between the corresponding nodes in the networks and, unless it is needed to obtain specific conclusions, the precise nature of nodes and edges is not required for their construction. We remark that although the data sets include the date of incorporation or establishment of offshore entities and intermediaries, they do not have specific information about any interaction dates between agents. For this reason edges in the networks could correspond to permanent or to occasional interactions, but it is not possible to know anything beyond the fact that an interaction between the corresponding agents occurred during their period of activity.

Additionally, we used the Paradise Papers data set from the same repository, solely to complement the jurisdiction analysis. This data set has the same structure as the Bahamas and Panama data sets. For jurisdiction analysis, we employed heterogeneous public data sets from the following institutions: Bank for International Settlements (BIS), Panama Banking Authority, Panama Capital Markets Authority, Central Bank of The Bahamas, Cayman Islands Monetary Authority, and Eastern Caribbean Central Bank. In the case of financial intermediaries, we extracted the location information from around 100 web sites of financial institutions with wealth management and private banking units. After manual processing, we obtained a data set that confirms the location in each jurisdiction.

8.3.2 Financial Network Structure

To understand how the large scale structure of the financial networks may evolve, we evaluate the role of three basic processes that in our opinion should be considered in a first approximation to their development. Notice that we refer here to the growth of the network determined by aggregate data up to a given moment and not to particular temporal states of the network activity. These processes are:

(P1) Once financial and non-financial intermediaries choose one or several jurisdictions with advantageous regulations, they come to those jurisdictions to develop their activities. No interaction with preexisting agents are necessary here.

(P2) New intermediaries appear that have direct interaction with preexisting intermediaries or with other new intermediaries that also arrive to initiate their activities. In both cases, they appeal to their clients (offshore entities).

(P3) Suspicious activities emerge in the network through the appearance of a small number of interactions (compared to the size of the network) that control authorities cannot easily detect. These interactions can occur between preexisting agents, or with new agents that due to their suspicious nature need to establish other additional connections to carry out their purposes.

Remark 8.1 Although in the development of a real network these processes are not expected to appear sequentially or even independently, their consideration is justified in our study because they constitute the smaller number of useful abstractions that isolate the essential aspects playing a distinct role in the behavior of these networks. Namely, the presence of a new or already established conglomerate of financial intermediaries offering services in their respective jurisdictions (P1), the appearance of new agents interacting in a mostly legal fashion with preexisting or upcoming members of the network (P2), and the emergence of corrupt interactions that occur basically in two ways: through established channels or involving new corrupt agents that, due to their role, require other additional interactions (P3). Clearly, more complicated processes in the network could be modeled using refinements and juxtapositions of these basic elements.

We employ these processes to study the plausibility that a simplified version of how the networks develop from a given moment can be obtained by assuming that from a number of established nodes (P1) the legal growth of the network is accounted solely through (P2) with the addition of new nodes with degree exactly one. Furthermore, suspicious interactions arise occasionally only through (P3) with the addition of either new edges alone, or of new nodes of degree larger than one. With these simplifications, the networks legal growth due to (P1) and (P2) can be understood as a result of the successive addition of new nodes with some preferential attachment scheme for their connections.

In fact, part of the motivation for these simplifications is that when an initial state is given by (P1) with no pre-existing edges between nodes, the resulting evolution can be described by the Barabási–Albert model [48, 49]. This model exhibits a scale-free property as a result of the interaction of two mechanisms: the continuous growth by successive addition of new nodes, and their preferential attachment to connect with already present nodes according to a probability proportional to the degree of the latter. Since in this case the degree distribution of the nodes obeys a power-law, a useful first test for these assumptions is to find how close the networks reflect a power-law behavior. It should not be drastically affected by the suspicious activities in (P3) if they represent a comparatively small proportion of all the interactions. The results of these tests for the studied networks are presented in Sect. 8.4.2, but it is important to notice here that a power-law behavior could appear only for a certain range of degrees depending on how restrictive are the simplified assumptions like the lack of initial edges in (P1).

8.3.3 A Community Detection Method

There exist several approaches [50–53] and methods to detect communities, e.g., the Louvain algorithm [54], the hierarchical clustering [55], the GN community structure algorithm [56, 57], the label propagation algorithm [58] and the Leiden algorithm [59, 60] to list a few. We present an elementary overview of the community detection problem to understand the network structure and the internal organization. If C denotes the set of all communities in a network, the *modularity* is defined by:

$$H = \frac{1}{2m} \sum_{c \in C} \left(e_c - \gamma \frac{K_c^2}{2m} \right), \tag{8.1}$$

where e_c is the number of edges in a community $c \in C$, m is the number of edges in the network, K_c is the sum of the degrees of the nodes appearing in that community, $\gamma > 0$ is a resolution parameter associated with the scale of the communities, and the term $K_c^2/2m$ corresponds to the expected number of edges [61]. We chose the Louvain algorithm to detect communities. This method optimizes the modularity in two phases that are repeated iteratively [54]. First, it assigns a different community to each node of the network, i.e., there are as many communities as there are nodes. In this phase, individual nodes are moved to the community that yields the largest increase in the quality function and stops when maximal modularity is attained. During the second aggregation phase, a network is created based on the partition obtained in the first phase whose nodes are now communities. Each community in this partition becomes a node in the aggregate network [54]. From there, we can identify how these communities reflect the Caribbean financial network structure and how intermediaries (financial and non-financial) and offshore entities in these networks interact with agents from other jurisdictions.

8.4   Results and Discussion

8.4.1   Jurisdiction Analysis

The purpose of this analysis is to identify the principal agents' jurisdiction. First, we used a data filter by intermediaries in the Bahamas Leaks and Panama Papers' data sets to identify the financial intermediaries' and non-financial intermediaries' localization. Second, we used other data filter by entities in the same data sets to identify the offshore entities' jurisdiction. The results show that the global financial intermediaries have preferred some jurisdictions in the Caribbean like Panama, The Bahamas, and the Cayman Islands to develop their activities, while a few global financial intermediaries prefer Turks and Caicos, Curaçao or St. Kitts and Nevis. On the other hand, some regional financial intermediaries prefer jurisdictions like Antigua and Barbuda, Grenada, and Saint Lucia (Fig. 8.1a). These results were different in the case of offshore entities because the jurisdictions with the largest agglomeration are smaller financial centers in the Caribbean like St. Kitts and Nevis, BVI, Barbados, and Aruba. However, the Bahamas, Panama, and the Cayman Islands still have an important participation in their financial tradition (Fig. 8.1a).

Another approximation to confirm the jurisdiction is through the identification of Caribbean offshore entities. We use only the data of the Caribbean jurisdiction from Bahamas Leaks and Panama Papers' data sets to explore the interaction characteristics between these entities and we complemented this analysis with the data set of Paradise Papers that shows interactions between different jurisdictions around the

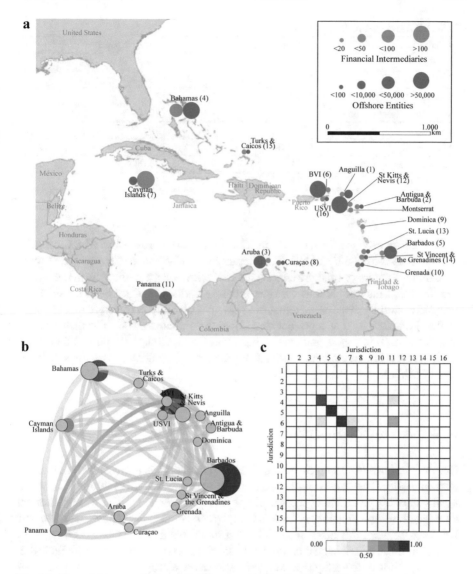

Fig. 8.1 Financial intermediaries and offshore entities by Caribbean jurisdictions. **a** In green the jurisdiction preference of financial intermediaries and in blue the jurisdiction preference of offshore entities. **b** Total of offshore entity interactions by Caribbean jurisdiction. **c** Weight of offshore entity interactions by Caribbean jurisdiction (numbers represent the name of the jurisdiction in Fig. 8.1a). Some offshore entities interact with entities in the same jurisdiction (self-loop), e.g.: Barbados = 40, 282; BVI = 40, 871; The Bahamas = 18, 245; Cayman Islands = 8, 871; Panama = 6, 108. *Note:* The offshore entity jurisdiction is the start point of the interaction, i.e., the origin of a directed edge

Table 8.2 Some networks measures

Variable	Bahamas leaks	Panama papers
Total nodes	219,856	559,433
Total edges	246,291	657,488
Average node degree	2.24	2.35
Avg. shortest path length	9.62	10.42
Diameter	29	39
Assortativity	−0.2404	−0.0521
Connected components	390	11,043
Detected communities	536	11,569

world, but we used the Caribbean data. The total of offshore entities with Caribbean jurisdictions is 585,200 and 127,672 interact in the Caribbean. Herewith, that more than three-quarters of offshore entities interact with other jurisdictions in the world. We identified the interactions between offshore entities by Caribbean jurisdictions (Fig. 8.1b) and the results show that the principal offshore entities are concentrated in five jurisdictions that represent Barbados, the Bahamas, BVI, Cayman Islands, and Panama. This conclusion may be based on the fact that these jurisdictions have the greatest interactions with entities from the same jurisdiction, i.e., the self-loops (see Fig. 8.1c).

8.4.2   Networks Analysis

Here we describe the complete curated financial networks in terms of their numerical characteristics. These networks have a low average node degree (see Table 8.2), which means that a big part of the agents connect to a single other agent. The average shortest path length for the Panama network is 9.62, which confirms the above, while for the Bahamas network this measure is 10.42, but this is not a representative value.

The number of connected components for the two data sets appear in Table 8.2. In general, it is noteworthy that both networks carry a proportionally small number of agents that interact with a high number of others that only have that connection, i.e., of degree one. For reference, we name those agents as central nodes. Finally, one of the principal characteristics of real networks is confirmed, not many agents are interconnected between themselves, to the contrary, most of the agents are not connected to the remaining majority. In several cases, especially in some communities the structure could resemble an egocentric network.

In view of the considerations explained in Sect. 8.3.2, we carried out a statistical analysis to study the behavior of the degree distribution of these networks. As illustrated on the left column of Fig. 8.2, the complementary Cumulative Distribution Function (CDF) of the degree centrality of each network shows that for some ranges

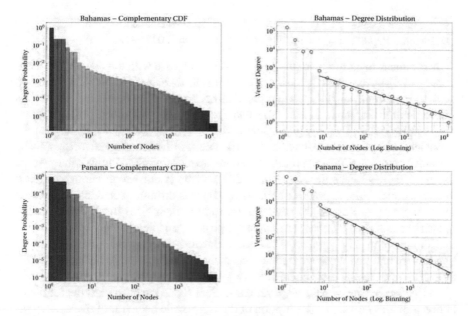

Fig. 8.2 Vertex degree histograms. First column: distributions of vertex degrees by number of nodes in logarithmic degree ranges for financial networks. Second column: analogous plots with logarithmic binning indicate that for some degree ranges the coarse behavior can be approximated by a power-law distribution. The corresponding scaling exponents are 1.602 and 2.066 for the Bahamas and Panama networks, respectively

between 10 and 10^3 nodes the degree distributions seem to obey an approximate power-law behavior $P(k) \sim k^{-\gamma}$. The right column of Fig. 8.2 contain plots of the corresponding probability distributions (PDF) using logarithmic binning, with a red line that corresponds to the best linear fitting using least squares. To corroborate these observations, we employed a more systematic power law fitting procedure following the methodology described in [62]. Briefly, the empirical CDFs of the networks were compared to the power law fitting given by the maximum likelihood estimator MLE for discrete distributions:

$$\hat{\gamma} \simeq 1 + n \left[\sum_{i=1}^{n} \ln \frac{x_i}{x_{\min} - \frac{1}{2}} \right]^{-1}, \qquad (8.2)$$

where n is the size of the sample, and $x_{\min}$ is the starting point or lower bound for the data which is chosen in order to provide the best possible fit. For the determination of this lower bound, a consecutive series of Kolmogorov–Smirnov (KS) tests is employed with $x_{\min}$ ranging from x_1 to x_n, and the final value of $x_{\min}$ is taken to be the one that minimizes the KS-statistic. The values found for the corresponding lower bound $x_{\min}$, scaling exponent $\hat{\gamma}$, and its error $(\delta\hat{\gamma})$ for the networks are:

⋄ Bahamas Leaks: $x_{\min} = 13$, $\hat{\gamma} = 1.60161$ ($\delta\hat{\gamma} = 0.02378$).
⋄ Panama Papers: $x_{\min} = 14$, $\hat{\gamma} = 2.06653$ ($\delta\hat{\gamma} = 0.01759$).

A goodness-of-fit test was also performed in each case with the power laws associated to the above values by means of the KS-statistic using a Monte-Carlo method with 1000 random samples. The corresponding p-values obtained from these tests where, for the Bahamas Leaks: 0.07 and for the Panama Papers: 0.113. From a conservative point of view it is considered that p-values above 0.1 indicate that the hypothesis of a power law behavior for the degree distribution of a network cannot be discarded, and this is the case for the Panama Papers network. Although the p-value for the Bahamas Leaks lies above 0.05 this statistic is not strong enough to reach the same conclusion. A possible explanation for this different behavior could come from the fact (see Sect. 8.4.4) that in proportion to their sizes the Bahamas Leaks network seems to carry a higher number of possibly suspicious interactions (which in addition are mostly confined to that same Bahamas jurisdiction) than the Panama Papers network, and those are interactions that are not expected to be accounted for by simple preferential attachment alone.

Indeed, according with the observations of Sect. 8.3.2, the power law degree distribution provides only a raw description of the large-scale behavior of these networks when the influence of processes of type (P3) are not significant in comparison to the total size of the network. Thus, in our case it should be expected that this method reflects more closely the structure of the Panama Papers network. Nevertheless, this method can still account for the negative assortativity found in both networks, which is due to the observed pattern that very highly connected agents tend to interact much more with very lowly connected ones, instead of interacting with other agents having also a high number of connections.

8.4.3 Community Analysis

We use a community detection algorithm to find communities inside each network beyond the given jurisdictions. From there, we can identify how these communities reflect the Caribbean financial network structure, and how intermediaries (financial and non-financial) and offshore entities in these networks interact with agents from other jurisdictions.

Our analysis reveals the community characteristics obtained from the Bahamas Leaks and Panama Papers data sets. The large scale distribution and complexity of these networks (approximately 400,000 nodes and 450,000 edges) present a sophisticated internal organization and compartmentalized structure into communities. We used the Louvain algorithm to identify how each node moves into a community and these sub-networks have been shown to have significant real-world meaning (Fig. 8.3). The distribution of communities permits to understand the structure of global financial networks, interactions with the Caribbean financial networks and their principal agents. These agents concentrate the greatest attention from authori-

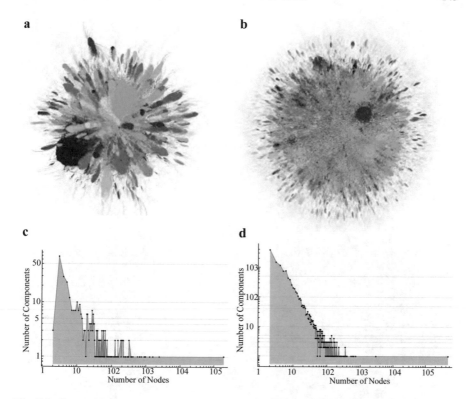

Fig. 8.3 Connected components of financial networks and their communities. The figures show the two largest connected components of the studied networks with their detected communities in different colors. First row: Bahamas Leaks network. Second row: Panama Papers network. Left column: largest connected components. Right column: second largest connected components. Note: Edge color represent the node color according to degree

ties and media but there exist other agents that are more relevant to understand the tax fraud, corruption, and money laundering activities. Nevertheless, we focused specifically on central nodes for two reasons: (1) we could confirm the possibility that some intermediaries show a high degree of interactions with different agents around the world and sometimes some offshore entities; and (2) this approach allows to broadly analyze the evolution of these agents in different periods, and particularly, how they are replaced.

The community structure analysis threw the following results. The Bahamas Leaks network (Fig. 8.4a) has 536 communities and the ten principal ones represent the 42.58% of total nodes (Fig. 8.4c). Two of the central nodes of these communities were Swiss bank subsidiaries in the Bahamas (Fig. 8.3a) and the others are corporate service firms (non-financial intermediaries) from the Bahamas (Fig. 8.3b). The Bahamas structure shows that interactions between certain agents are not only with the central node but with other agents creating sub-networks (see Fig. 8.3a). This behavior of separate agents is a result of the type of interactions between agents and

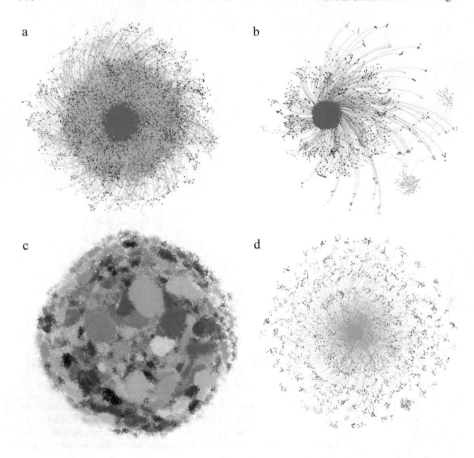

Fig. 8.4 Financial networks communities and connected components. First row: complete networks, including detected communities and interactions among offshore entities, intermediaries and officers. Second row: plots of number of components versus number of nodes (per component) for each network. Left column: Bahamas Leaks Network. Right column: Panama Papers Network

the suspicious activity rules. Some agents interact with only one agent or with a small group of agents, while the central nodes interact with these agents and with agents of some other communities. As a result, some communities are disconnected in the Bahamas network.

For the Panama Papers network, we obtained the following results (Fig. 8.4b). The ten largest communities of around 11,500 communities represent the 15.43% of total nodes (Fig. 8.4d). The main communities in the Panama Papers, as well as in the case of the Bahamas Leaks, have a large number of interactions between the nodes of each community because central nodes of these communities are mainly financial intermediaries or, in their absence, fiduciary or legal intermediaries. These intermediaries show thousands of interactions with other agents, especially offshore entities and officers that interact only with those central nodes (Fig. 8.3). Although due to the

Table 8.3 Evolution of the Panama Papers Network

Period	Nodes	Edges	Communities
1970–1994[a]	47,814	46,539	4,425
1995–1999	63,775	67,258	3,578
2000–2004	95,945	105,712	3,632
2005–2009	117,168	129,849	3,578
2010–2014	80,666	88,655	2,897
2015–2017[b]	8,450	8,277	666

- [a]Period with interactions before 1990
- [b]Period until data available

size and the structure of the networks many detected communities correspond precisely to connected components, large connected components carry a great number of detectable individual communities inside themselves (e.g. the largest component of Panama in Fig. 8.3c).

8.4.4 Dynamical Structure Analysis

To illustrate the structural evolution of networks, we employ ensembles of time-dependent networks extracted from the Bahamas Leaks and the Panama Papers data sets. We used the establishment date data of offshore entities or intermediaries to identify the networks' evolution. Additionally, we divided the data set into periods of five years of the whole temporal range considered (see Table 8.3). The Panama Papers network results confirmed the role of some central nodes and their evolution. A useful example is Mossack Fonseca and its offices around the world (Fig. 8.5). The first years with its office in Geneva, later with its office in London and this century with its office in Singapore and connections with its offices in London, Panama, and Lima as central nodes of different communities. These central nodes interacted with offshore entities, and officers in Caribbean jurisdictions. However, during the decade 2000–2010, new central nodes have appeared of entities and intermediaries in Caribbean jurisdictions and business in the Asia-Pacific, principally, with jurisdictions in Honk Kong, Singapore, and Taiwan.

Although the Panama Papers are a leak of Mossack Fonseca, they lost their position as the principal-agent while other intermediaries achieved better performance. We may analyze some examples. By the first period, Mossack Fonseca Geneva was the central node of the biggest community with 4.82% of total network interactions (Fig. 8.5a). However, the Geneva office of Mossack Fonseca was never again relevant in the largest communities because other intermediaries won relevance, including Mossak Fonseca offices in different jurisdictions.

Another example is the evolution of Mossack Fonseca Singapore that created an ecosystem in this jurisdiction with other intermediaries in Hong Kong and Taiwan with a community that had 2.23% of total interactions for the 2000–2004 period (Fig. 8.5c). Nevertheless, other central nodes (of Southeastern Asia origin) and their communities were more relevant in the number of interactions than Mossack Fonseca with 3.28% and 2.61%, respectively. The community structure analysis confirms the evolution of the Panama Papers network as a process of global interactions where the Caribbean jurisdictions are part of a worldwide network that hid behind an evolving financial network (Fig. 8.5).

Additionally, with the results of community structure analysis we identified the business evolution in tax havens, and particularly, the size of largest communities in the different periods. In other words, we could confront these results with known aspects of the relevance of financial centers in Swiss cities and London in the UK during the first periods, as well as changes in tax laws in Switzerland and the UK during the last years of the 1990s. Moreover, the new law framework that helped to consolidate the Singapore and Hong Kong financial systems during the 2000s opened the possibility to create new offshore services in these jurisdictions, as well as the economic evolution in China.

To summarize, we found that communities associated with central nodes in the Panama Papers elicited change from financial intermediaries to non-financial intermediaries, as a result of stricter control and regulation of financial intermediaries than of professional services. As expected, we found that the number of financial intermediaries decreased over time, and the non-financial intermediaries (legal, accounting, and management) increased their participation in financial networks. Furthermore, with this temporal decomposition of the Panama Papers network we confirmed that, in relation to the establishment and liquidation of agents, the most frequent was of offshore entities, to a lesser extend of non-financial intermediaries, and to a much lesser one of financial intermediaries (Table 8.3).

With these results, we could identify central nodes, more visible agents to authorities and media, and their evolution, i.e., the data analysis confirmed some general characteristics of the financial crime problem (see Table 8.3). But some measures indicate that it is necessary to use other tools to identify suspicious groups. One of them is the negative assortativity of the networks because it is a representation of a characteristic that may differ for each node in the graph [63]. The assortativity for the Bahamas network is −0.2404, and for the Panama network it is −0.0521. In this context, a network that is non-assortative, i.e., with negative assortativity, may comprise agents that are themselves highly assortative but others lowly assortative because an observed particularity of financial crime networks is that some agents do not have the interest to connect with other high degree agents, only in some situations or through another agent (see for example Fig. 8.5d or f).

Finally, the boom of offshore activities in The Bahamas during the 1990s presented a specific structural evolution between 1990 to 2001. We divided the data set into periods of two years of the first years of the temporal range considered (see Table 8.4).

Similar to Panama Papers, the results confirmed the role of some central nodes. In this case, the financial intermediaries play an important role in this network during the boom of activities (Fig. 8.6). After 2000, the stability of intermediaries is a result of the strict regulation process in The Bahamas that reduces the proliferation of

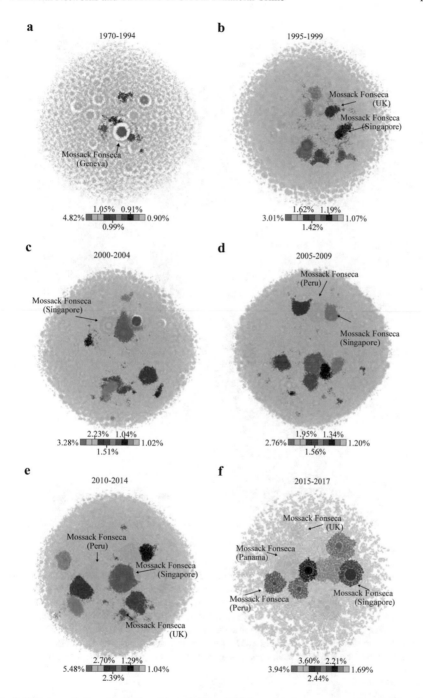

Fig. 8.5 Temporal decomposition and evolution of 10 largest panama papers communities. For each period the color scales reflect the ranking of the largest detected communities ordered by the percentage of their size in the entire network

a   b

■ 1990-1991   □ 1992-1993

c

■ 1994-1995   □ 1996-1997

□ 1998-1999   ■ 2000-2001

Fig. 8.6 Comparison of temporal decompositions of the new Bahamas Network interactions during the boom of activities of the 1990s. The network nodes are grouped by communities, and the size of the three graphs have the same scale. Notice that for the first two periods the amount of new interactions is larger than the last one where colors appear inverted. The inversion is due to the fact that for good contrast gray was used for the larger network, while green for the smaller one, but in the fisrt two graphs the larger network appears at the final period, i.e. the networks arw growing, while for the last graph the larger network appears at the starting period, i.e. the network is shrinking

Table 8.4 The boom of activities in the Bahamas leaks network

Period	Nodes	Edges	Communities
1990–1991	8,645	8,446	222
1992–1993	13,658	13,417	255
1994–1995	23,544	23,256	315
1996–1997	35,742	35,732	351
1998–1999	34,884	34,582	406
2000–2001	26,309	25,980	537

financial intermediaries and non-financial intermediaries. Additionally, we observed the establishment of banks and financial institutions of principal global financial centers, which develop their activities since the 1990s. The intermediaries interacted, especially, with offshore entities and officers in the Bahamas jurisdiction. However, during the 1990s, in the Bahamas were created approximately 120,000 offshore entities and intermediaries that needed additional officers to attend the new customers and interactions (see Table 8.4).

8.5 Conclusions

We have studied the peculiarities of suspicious activities and financial crimes in large-scale financial networks. Because financial networks present some structural properties common to other complex networks that affect the identification of suspicious activities, our procedures start by integrating data analysis and community structure methodologies to identify and understand some elements of this problem.

For this purpose, we analyzed the structure of financial networks for thirty years, using information from the International Consortium of Investigative Journalists. These data sets allowed us to obtain details about the evolution of financial networks and characteristics of tax havens, offshore entities and their relations with other jurisdictions. We observed that some general aspects of suspicious activities in large-scale networks can be detected with a characterization of their elements using network structure methodologies. But this strategy needs complementary information to identify the details of these phenomena because financial criminals in global or local scenarios develop their activities in smaller groups.

Our numerical analysis for the degree distributions of these networks determined a range of degree values for which their frequency distribution exhibit more closely a power law behavior. Indeed, for the Panama Papers network the hypothesis that above an obtained lower degree bound the distribution obeys exactly a power law cannot be statistically discarded, while for the Bahamas Leaks network additional considerations seem to be required to explain its overall structure given that the statistics were shown to be a bit weaker.

We indicated useful perspectives to understand the dynamics of problems as tax fraud, corruption, and money laundering activities in financial networks. The analysis carried out so far has focused on visible agents and interaction facilitators as well as financial intermediaries. This methodology can help to shed light on many other agents that go unnoticed in these large-scale financial networks that grow in different ways because it describes the emergence of locality and the characteristics of their interactions.

Several offshore entities have their jurisdiction in the Caribbean financial centers, but only a quarter of them operate with other Caribbean entities, i.e., these entities do not always have their interactions with intermediaries in the Caribbean jurisdictions. Thus, financial crime is not confined to tax havens but also different agents use these jurisdictions to arbitrage taxes and for money laundering as part of global interactions. Our results can facilitate a new approach to financial crime analysis and the relationship between different groups of agents in financial networks. Additionally, it can improve our understanding of the structure of corruption and money laundering, and finally, it can facilitate the exploration of unnoticed small groups which are the basis of global financial crime.

For future work, a natural next step is to employ other mathematical tools like geometric and topological notions that can be useful in this setting. More concretely, from the topological side, a characterization can be made of the suspicious interactions that indicate corrupt activities in terms their abstract topological properties. An example would be the presence of certain cycles (or loops) in the financial networks. This would require a more mathematical formalization of what those suspicious interactions are as to allow the use of homological techniques and invariants (e.g. Betti numbers) to measure and evaluate the connectivity structure of the networks. From the geometrical counterpart it is expected that the role of notions like discrete curvature can be adapted to provide a clearer picture of the studied interactions in these networks and possibly also aid in the detection of some kinds of suspicious activities.

Acknowledgements The authors would like to thank the anonymous reviewer of this chapter for the useful comments and suggestions to this manuscript and to the previous versions presented in NetSci 2019 and Lanet 2019.

References

1. UNODC (2018) Money-laundering and globalization. https://www.unodc.org/unodc/en/money-laundering/globalization.html
2. Ribeiro HV, Alves LGA, Martins AF, Lenzi EK, Perc M (2018) J Complex Netw 6(6):989. https://doi.org/10.1093/comnet/cny002
3. Calderón DJG (2018) Análisis Político 31(92):180. https://doi.org/10.15446/anpol.v31n92.71106
4. Niklander S, Soto R, Crawford B, de la Barra CL (2016) E. Olguín. In: Stephanidis C (ed) HCI international 2016 - Posters' extended abstracts. Springer International Publishing, Cham, pp 63–66

5. Zhan JV (2012) Crime. Law Soc Change 58(2):93. https://doi.org/10.1007/s10611-012-9379-9
6. Wachs J, Yasseri T, Lengyel B, Kertész J (2019) R Soc Open Sci 6(4):182103 (2019). https://doi.org/10.1098/rsos.182103
7. Nese A, Troisi R (2019) Trends Organ Crime 22(3):298. https://doi.org/10.1007/s12117-018-9349-4
8. Morris SD (2013) Trends Organ Crime 16(2):195. https://doi.org/10.1007/s12117-013-9191-7
9. Luna-Pla I, Nicolás-Carlock JR (2020) Appl Netw Sci 5(1):13. https://doi.org/10.1007/s41109-020-00258-2
10. Chayes S (2016) The structure of corruption: a systemic analysis using Eurasian cases. Carnegie Endowment for International Peace, Washington
11. Cartier-Bresson J (1997) Polit Stud 45(3):463. https://doi.org/10.1111/1467-9248.00091
12. Kanin DB (2003) Int Polit 40(4):491. https://doi.org/10.1057/palgrave.ip.8800038
13. Gerring J, Thacker SC (2004) British J Polit Sci 34(2):295. https://doi.org/10.1017/S0007123404000067
14. Johnston M (2013) Corruption, contention, and reform: the power of deep democratization. Cambridge University Press, Cambridge. https://doi.org/10.1017/CBO9781139540957
15. Pring C (2017) People and corruption: citizens voices from around the world. Working paper, Transparency International
16. Kuris G (2015) Policy Soc 34(2):125. https://doi.org/10.1016/j.polsoc.2015.04.003
17. Guiso L, Sapienza P, Zingales L (2011) In: Benhabib J, Bisin A, Jackson MO (eds) Handbook of social economics, vol 1. North-Holland, pp 417–480. https://doi.org/10.1016/B978-0-444-53187-2.00010-3
18. Warburton J (2013) Corruption and anti-corruption. In: Chapter Corruption as a social process: from dyads to networks. ANU Press, Canberra, pp 221–237. https://doi.org/10.22459/CAC.03.2013.13
19. Celentani M, Ganuza JJ (2002) Ann Oper Res 109(1):293. https://doi.org/10.1023/A:1016364505439
20. Venard B, Hanafi M (2008) J Busin Ethics 81(2):481. https://doi.org/10.1007/s10551-007-9519-9
21. Khwaja A, Mian A (2011) Ann Rev Econ 3. https://doi.org/10.1146/annurev-economics-061109-080310
22. Campos N, Engel E, Fischer RD, Galetovic A (2019) Renegotiations and corruption in infrastructure: the Odebrecht case. "Marco Fanno" Working Papers 0230, Dipartimento di Scienze Economiche "Marco Fanno". https://ideas.repec.org/p/pad/wpaper/0230.html
23. Bjørnskov C (2003) Corruption and social capital. Working papers 03-13, University of Aarhus, Aarhus School of Business, Department of Economics
24. Harris D (2007) Bonding social capital and corruption: a cross-national empirical analysis. Environmental economy and policy research working papers 27.2007, University of Cambridge, Department of Land Economics (2007)
25. Neanidis KC, Rana MP, Blackburn K (2017) Ann Financ 13(3):273. https://doi.org/10.1007/s10436-017-0299-7
26. Neumann M, Elsenbroich C (2017) Trends Organ Crime 20(1):1. https://doi.org/10.1007/s12117-016-9294-z
27. Ardizzi G, Franceschis PD, Giammatteo M (2018) Int Rev Law Econ 56:105. https://doi.org/10.1016/j.irle.2018.08.001
28. Loayza N, Villa E, Misas M (2019) J Econ Behav Organ 159:442. https://doi.org/10.1016/j.jebo.2017.10.002
29. Walker J (1999) J Money Laund Control 3(1):25. https://doi.org/10.1108/eb027208
30. Masciandaro D, Takáts E, Unger B (2007) Black finance. Edward Elgar Publishing, Cheltenham
31. Schneider F (2010) Public Choice 144(3/4):473. https://doi.org/10.1007/s11127-010-9676-8
32. McCarthy KJ, van Santen P, Fiedler I (2015) Int Rev Law Econ 43:148 (2015). https://doi.org/10.1016/j.irle.2014.04.006

33. Imanpour M, Rosenkranz S, Westbrock B, Unger B, Ferwerda J (2019) Int Rev Law Econ 60. https://doi.org/10.1016/j.irle.2019.105856
34. Walker J, Unger B (2009) Rev Law Econ 5. https://doi.org/10.2202/1555-5879.1418
35. Dreżewski R, Sepielak J, Filipkowski W (2015) Inf Sci 295:18 (2015). https://doi.org/10.1016/j.ins.2014.10.015
36. Colladon AF, Remondi E (2017) Expert Syst Appl 67:49. https://doi.org/10.1016/j.eswa.2016.09.029
37. Vaithilingam S, Nair M (2009) Res Int Busin Financ 23(1):18. https://doi.org/10.1016/j.ribaf.2008.03.003
38. García-Bedoya O, Granados O, Cardozo J (2020) J Money Laund Control. https://doi.org/10.1108/JMLC-04-2020-0033
39. Eden L (2009) Chapter, Taxes, transfer pricing, and the multinational enterprise. In: The Oxford handbook of international business, 2 ed. Oxford University Press. https://doi.org/10.1093/oxfordhb/9780199234257.003.0021
40. Frunza MC (2016) In: Frunza MC (ed) Introduction to the theories and varieties of modern crime in financial markets. Academic, San Diego, pp 153–164. https://doi.org/10.1016/B978-0-12-801221-5.00011-7
41. Jones C, Temouri Y, Cobham A (2018) J World Busin 53(2):177. https://doi.org/10.1016/j.jwb.2017.10.004
42. Schwarz P (2011) Int Rev Law Econ 31(1):37. https://doi.org/10.1016/j.irle.2010.12.001
43. Menkhoff L, Miethe J (2019) J Publ Econ 176:53. https://doi.org/10.1016/j.jpubeco.2019.06.003
44. Schweitzer F, Fagiolo G, Sornette D, Vega-Redondo F, Vespignani A, White DR (2009) Science 325(5939):422. https://doi.org/10.1126/science.1173644
45. Newman M (2018) Networks: an introduction, 2nd edn. OUP Oxford, Oxford
46. Shaxson N (2011) Treasure Islands: tax havens and the men who stole the world. St. Martin's Griffin
47. TIC of Investigative Journalists (2017) Offshore leaks database. https://offshoreleaks.icij.org/
48. Barabási AL, Albert R (1999) Science 286(5439):509. https://doi.org/10.1126/science.286.5439.509
49. Albert R, Barabási AL (2001) Rev Modern Phys 74. https://doi.org/10.1103/RevModPhys.74.47
50. Danon L, Díaz-Guilera A, Duch J, Arenas A (2005) J Stat Mech: Theory Exper 2005(09):P09008. https://doi.org/10.1088/1742-5468/2005/09/p09008
51. Fortunato S, Barthélemy M (2007) Proc Natl Acad Sci 104(1):36. https://doi.org/10.1073/pnas.0605965104
52. Fortunato S (2010) Phys Rep 486(3):75. https://doi.org/10.1016/j.physrep.2009.11.002
53. Newman MEJ, Girvan M (2004) Phys Rev E 69. https://doi.org/10.1103/PhysRevE.69.026113
54. Blondel VD, Guillaume JL, Lambiotte R, Lefebvre E (2008) J Stat Mech: Theory Exper 2008(10):P10008 (2008). https://doi.org/10.1088/1742-5468/2008/10/p10008
55. Lancichinetti A, Fortunato S, Radicchi F (2008) Statistical, nonlinear, and soft matter physics. Phys Rev E 78. https://doi.org/10.1103/PhysRevE.78.046110
56. Girvan M, Newman MEJ (2002) Proc Natl Acad Sci 99(12):7821. https://doi.org/10.1073/pnas.122653799
57. Radicchi F, Castellano C, Cecconi F, Loreto V, Parisi D (2004) Proc Natl Acad Sci 101(9):2658. https://doi.org/10.1073/pnas.0400054101
58. Raghavan UN, Albert R, Kumara S (2007) Phys Rev E 76. https://doi.org/10.1103/PhysRevE.76.036106
59. Traag VA (2015) Phys Rev E 92. https://doi.org/10.1103/PhysRevE.92.032801
60. Traag VA, Waltman L, van Eck NJ (2019) Sci Rep 9(1):5233. https://doi.org/10.1038/s41598-019-41695-z
61. Reichardt J, Bornholdt S (2006) Phys Rev E 74. https://doi.org/10.1103/PhysRevE.74.016110
62. Clauset A, Shalizi CR, Newman M (2009) SIAM Rev 51(4):661. https://doi.org/10.1137/070710111
63. Noldus R, Van Mieghem P (2015) J Complex Netw 3(4):507. https://doi.org/10.1093/comnet/cnv005

Chapter 9
Corruptomics

José R. Nicolás-Carlock and Issa Luna-Pla

> *Injustice anywhere is a threat to justice everywhere. We are caught in an inescapable network of mutuality, tied in a single garment of destiny. Whatever affects one directly, affects all indirectly.*
> *–Martin Luther King Jr.*

Abstract Corruption studies must evolve to match the complexity of the modern world. Here, we present three main problems in corruption analysis that need to be address: the complexity of the corruption phenomenon itself and its context, the complexity of the analytical description, and the complexity of the perspectives that different disciplines bring to the table. In this regard, we argue that the interdisciplinary framework of complex systems and network science represents a promising analytical approach to move forward in this endeavor. Furthermore, current research efforts in this direction indicate the dawn of a new interdisciplinary discipline for corruption studies.

Corruption is one of the most prominent global policy challenges of the 21st century. During the last few decades, corruption not only has been extensively addressed in the public policy arena but also, it has become a very active academic research field [19, 21]. As an academic subject, corruption is mostly regarded as a fundamental societal problem that researchers from diverse disciplinary traditions aim to address along four main interdependent axes: conceptualizations and definitions, measuring methods and techniques, modelling of causes and consequences, and control or tackling strategies. However, although relevant advances have been made, the challenge to design theoretical and technical frameworks that are able to handle the complex-

J. R. Nicolás-Carlock (✉) · I. Luna-Pla
Institute of Legal Research, National Autonomous University of Mexico,
Mexico City, Mexico
e-mail: jnicolas@unam.mx

I. Luna-Pla
e-mail: ilunapla@unam.mx

ity of this phenomenon remains highly contested [14]. This is due to three main problems in corruption analysis: (I) the complexity of the nature of the phenomenon itself and its context, (II) the complexity of the analytical description, and (III) the complexity of the different perspectives that each discipline brings to the table.

I. The complexity of the nature of the phenomenon itself and its context. Human beings are complex and corruption is inherently hard to tackle due to the complex nature of human behavior. As any other human activity, corruption occurs within the intricate structure and dynamics of the social, economic and political systems of society. As such, corrupt behavior manifests as a non-separable activity that tends to be hidden and interwoven within the multiple activities that could be deem as non-corrupt for a given setting, from the micro group dynamics that take place within the structure of government institutions, small or big corporations, and civil life, to the macro interactions that take place among them. In addition, the systems that characterize our complex societies are not independent of each other but constitute a system of systems whose structure and dynamics are always evolving in response to changes in the corresponding social and regulation context. Notably, global dynamics have radically changed in the last few years. The flow of people, materials, and information have remarkably increased the levels of interactions at different spatial and administrative scales and consequently, the interdependencies among social, economic and political systems have become stronger. Our heterogeneous world is more connected than ever. This not only represents great cooperation opportunities for development but also highly systemic threats at regional and global scales, what happens in one place or sector can have effects on other seemingly unrelated ones, putting the analysis of corruption on a whole another level [1, 12].

II. The complexity of the analytical description. The analysis of corruption is complex and has evolved over a long period of time. Nowadays, this analysis is done under an international consensus that has put individual behavior and indiscretions at the center of modern corruption thinking [14]. Values, norms and ideas, that still play an important role in our understanding of what is acceptable or not in society, vary from place to place, and from time to time, making an objective understanding of moral and ethical issues a great challenge [5]. In an effort to be more objective, the analysis of corruption has opted for a more pragmatic and empirical approach.

On the conceptual dimension, contemporary thinking is dominated by four main approaches: legal definitions, the "abuse of entrusted power" criterion, economic or business-oriented, and the so-called "legal" corruption. Among these, the first one has become the most popular since its adoption from Transparency International, the Organization for Economic Cooperation and Development, the World Bank and the International Monetary Fund. However, the subjective elements embedded in this definition, such as what constitutes "abuse" and its bias towards the public sector, make it object of continuous critic and debate. On the measuring front, aggregated indices rise as the most popular as they have provided an overall and broad picture of global corruption. These still face criticism due to changes in their methodology or the perception/experience-based analysis since that makes the interpretation or comparison of results difficult. This has lead to the creation of other promising indicators that address more specific matters, as well as more sophisticated approaches based on proxies to corruption that have the capacity to describe macro features from micro data. On the modelling challenge, the consensus is low on what are the general causes of corruption. The different models based on structural forces, rational agents, principals or discretionary criteria provide insight for some settings but not all, therefore, these too are not free from debate due to the emergence of interesting collective dilemmas and the clash of social realities across the globe. On the control arena, the strategies to tackle corruption are not universal and, in direct relation with the causes of the problem at hand, strategies depend on the specific context. National regulations and international instruments aim to solve this problem but their results are not clear and there's no catalogue of solutions that work for all places at all times [14].

III. The complexity of different disciplinary perspectives. As multifaceted as the phenomenon is, corruption analysis has been greatly enriched by the insights of researchers from different disciplines and schools of thought. To this day, these are mostly from political science, economy, sociology, anthropology, or law. In the effort to come up with a general interdisciplinary corruption theory, these researchers have to deal not only with the complex nature of phenomenon itself, the inherent heterogeneity and complexity of the systems where it takes place, or the subtleties of the four dimensions that comprise modern analysis, but also, with the clash of different ideas and perspectives about social reality that each discipline brings to the table [7]. This is a complex scenario where consensus on fundamental aspects is hard to achieve and, therefore, a general interdisciplinary theory of corruption has remained elusive [14].

The challenge is great and in order to move forward, it's imperative to adopt new ideas and perspectives that enable us to handle and embrace the complexity of our world instead of avoid it. Here, complexity or complex systems science represent a promising approach in this endeavour:

- First, we must understand that the complexity of society is nothing but the result of our collective doing as we become and create the systems that shape our social, economic and political environments. In other words, we not only are but also create the complex networks in which corruption takes place. Therefore, our connections and interactions at the different scales, sectors and regions keep valuable information that can be tapped, modelled and studied in order to understand the principles that govern such systems and give us the opportunity to develop strategies and interventions tailored to the structure and dynamics of the problem at hand. In that regard, complexity science is an interdisciplinary approach to the study of collective phenomena in natural, social and technical systems that has been successful in the analysis of the structure and dynamics of such systems in terms of the relationships among their parts and their environment [2, 3]. The main idea in complex systems is that a collection of interacting components behaves in way not predicted by the components in isolation or disconnected. Interactions and dependencies matter more than the nature of the parts. Therein, the collection of interacting parts is best understood as a whole, rather than disconnected. As such, when corruption behavior manifest as a non-separable activity that tends to be hidden and intertwined among the multiple activities that occur within the structure and dynamics of social, economic, political and technical systems, then, activities that could be deemed as corrupt are best understood systemically, this is, from the collective behavior and features of connected individuals or organizations acting as a whole, rather than from their particular characteristics in isolation.

- Second, given the multifaceted nature of corruption, the importance of a comprehensive and general analytical framework—not necessarily universal but that unifies the different disciplinary perspectives—and that encompasses proper conceptualizations and definitions of corrupt practice cannot be overstated, given that definitions determine what gets modelled and what researchers look for in data, in such a way that unsuited definitions can cause misleading measurements, mistaken interpretations of causes and consequences, and ultimately inappropriate policy suggestions. Achieving a unified framework seems daunting, but here again, complexity science presents itself as an interesting and relevant example on this matter, as a discipline that on its own is dealing with a similar challenge [16]. For that, let us recall that complexity science is an interdisciplinary approach to the study of natural, social and technical systems that has created an still-evolving analytical framework by drawing concepts and tools from disciplines such as physics, chemistry, biology, ecology, sociology, mathematics and computer science [18]. In this way, complex systems are often studied in terms of networks, self-organization, evolution, non-linearity, scaling, and emergence. Remarkably, this framework has been applied to different systems and settings leading to relevant insights into crime, terrorism, war, disease spreading, financial markets, democracy and other social subjects [4, 8, 13, 15]. In the case of corruption, the concepts and tools of complexity and networks have been applied to tackle specific matters, such as the conceptualization of corruption as a networked phenomenon [22], measuring and modelling of political corruption [6, 20, 23], corruption in public procurement and corruption scandals [9, 10, 17, 24], and ways to identify, counteract and control phenomena such as cartel formation, money laundering and tax evasion by means of data and artificial intelligence [11, 25, 26].

- Lastly, anti-corruption is not an endeavor of isolated and disconnected individuals. The efforts made in the public policy arena have shown that groups of people, private institutions, and governments must learn to cooperate and work purposely in order for any strategy to be effective and successful. In this world, we are not only connected at multiple scales but also we are interdependent members of this highly complex system known as society: whatever affects one directly, affects all indirectly. Complex corruption networks are tackled with complex anti-corruption networks, and corruption studies and anti-corruption strategies must evolve to match the complexity of our reality.

Modern technological advances by themselves are not sufficient to suggest that we have the upper-hand in the fight against corruption since, in the same way those emerging technologies allow for greater transparency, cooperation and development, they could also allow for more sophisticated corruption mechanisms and challenges. It comes down to us to make the best out of the modern tools and achievements obtained so far in the long history of corruption analysis and move forward. The research done so far to address corruption, such as the one presented all along the chapters of this book, constitute an effort in this direction, as well as example of the scope, possibilities, and potential of a new approach to corruption analysis that is open to perspectives, methods and disciplines that go beyond the traditional schools of thought in order to give place to something different and hopefully useful.

We might not be any closer to a corruption free era, but we are positive that we are at the dawn of a new paradigm in corruption and anti-corruption studies that could take us closer to that goal. Let us consider this an opportunity to create a new discipline that embraces the complexities of the phenomenon, that takes into account the structure and dynamics of the networks that shape our societies, that takes advantage of technology, data and empirical evidence, that pulls insight from the full spectrum of the prism of interdisciplinary science, and finally, that dares to take corruption studies closer to an exact science. Let us consider and know henceforth this new discipline as *corruptomics*: corruption analysis for the 21st century.

References

1. Balsa-Barreiro J, Vié A, Morales AJ, Cebrián M (2020) Deglobalization in a hyper-connected world. Palgrave Commun 6(1):1–4
2. Bar-Yam Y (2004) Making things work: solving complex problems in a complex world. Knowl Ind
3. Barabási AL (2016) Network science. Cambridge University Press, Cambridge
4. Caldarelli G (2020) A perspective on complexity and networks science. J Phys: Complex 1(2):021001
5. Capraro V, Perc M (2018) Grand challenges in social physics: in pursuit of moral behavior. Front Phys 6:107
6. Colliri T, Zhao L (2019) Analyzing the bills-voting dynamics and predicting corruption-convictions among Brazilian congressmen through temporal networks. Sci Rep 9(1):16754
7. De Graaf G, Von Maravic P, Waagenar P (2010) The good cause: theoretical perspectives on corruption. Verlag Barbara Budrich
8. Eliassi-Rad T, Farrell H, Garcia D, Lewandowsky S, Palacios P, Ross D, Sornette D, Thébault K, Wiesner K (2020) What science can do for democracy: a complexity science approach. Human Soc Sci Commun 7(1):1–4
9. Fazekas M, Wachs J (2020) Corruption and the network structure of public contracting markets across government change. Polit Govern 8(2):153–166
10. Fierăscu SI (2017) The networked phenomenon of state capture. PhD thesis, Central European University
11. Garcia-Bedoya O, Granados O, Burgos JC (2020) Ai against money laundering networks: the Colombian case. J Money Laund Control
12. Helbing D (2013) Globally networked risks and how to respond. Nature 497(7447):51–59
13. Helbing D, Brockmann D, Chadefaux T, Donnay K, Blanke U, Woolley-Meza O, Moussaid M, Johansson A, Krause J, Schutte S et al (2015) Saving human lives: what complexity science and information systems can contribute. J Stat Phys 158(3):735–781
14. Hough D (2017) Analysing corruption. Agenda Publishing Limited
15. Kertész J, Wachs J (2020) Complexity science approach to economic crime. Nat Rev Phys
16. Ladyman J, Wiesner K (2020) What is a complex system? Yale University Press
17. Luna-Pla I, Nicolás-Carlock J (2020) Corruption and complexity: a scientific framework for the analysis of corruption networks. Appl Netw Sci 5(1):13

18. Mitchell M (2009) Complexity: a guided tour. Oxford University Press, Oxford
19. Mungiu-Pippidi A, Heywood PM (2020) A research agenda for studies of corruption. Edward Elgar Publishing
20. Ribeiro HV, Alves LGA, Martins AF, Lenzi EK, Perc M (2018) The dynamical structure of political corruption networks. J Complex Netw 6(6):989–1003
21. Rose-Ackerman S, Palifka BJ (2016) Corruption and government: causes, consequences, and reform. Cambridge University Press
22. Slingerland W (2018) Network corruption: when social capital becomes corrupted. Eleven International Publishing
23. Solimine PC (2020) Political corruption and the congestion of controllability in social networks. Appl Netw Sci 5(1):1–20
24. Wachs J, Fazekas M, Kertész J (2020) Corruption risk in contracting markets: a network science perspective. Int J Data Sci Anal
25. Wachs J, Kertész J (2019) A network approach to cartel detection in public auction markets. Sci Rep 9(1):1–10
26. Zumaya M, Guerrero R, Islas E, Pineda O, Gershenson C, Iñiguez G, Pineda C (2021) Identifying tax evasion in Mexico with tools from network science and machine learning. arXiv preprint arXiv:2104.13353

Printed in the United States
by Baker & Taylor Publisher Services